VISUALITY/MATERIALITY

Visuality/Materiality

Images, Objects and Practices

Edited by

GILLIAN ROSE
The Open University, UK
DIVYA P. TOLIA-KELLY
University of Durham, UK

Routledge
Taylor & Francis Group

LONDON AND NEW YORK

First published 2012 by Ashgate Publishing

2 Park Square, Milton Park, Abingdon, Oxon OX14 4RN
711 Third Avenue, New York, NY 10017, USA

Routledge is an imprint of the Taylor & Francis Group, an informa business

First issued in paperback 2016

British Library Cataloguing in Publication Data
Visuality/materiality : images, objects and practices.
   1. Visual communication. 2. Visual literacy.
   3. Geographical perception. 4. Human ecology.
   I. Rose, Gillian, 1962- II. Tolia-Kelly, Divya Praful.
   304.2-dc23

Library of Congress Cataloging-in-Publication Data
Visuality/ materiality : images, objects and practices / by Gillian Rose and Divya P. Tolia-Kelly., [editors].
   p. cm.
   Includes bibliographical references and index.
   ISBN 9781409412229 (hardback) -- ISBN 9781409412236
   (ebook) 1. Material culture. 2. Visual perception. 3. Visual
   communicaiton. I. Rose, Gillian. II. Tolia-Kelly, Divya Praful.
   GN406.V57 2012
   302.2'3--dc23

2011037642

ISBN 978-1-4094-1222-9 (hbk)
ISBN 978-1-138-25251-6 (pbk)

Contents

List of Figures

About the Contributors

Editors

Gillian Rose is a Professor in Cultural Geography at the Open University. She is a significant author within the field of visual culture. Her particular interest is in considering visuality as a kind of practice, done by human subjects in collaboration with different kinds of objects and technologies. Her current research includes working with Architects on UK Research Council funded project *Architectural Atmospheres* which follows from the research project *Urban Aesthetics*. She is the author of *Visual Methodologies* (Palgrave, 2001) *and Doing Family Photography: The Domestic, the Public and the Politics of Sentiment* (Ashgate, 2010).

Divya P. Tolia-Kelly is a Reader in Geography at Durham University, UK. Her research has focused on visual cultures, material cultures, landscape and race-memory. In collaborations with landscape artists Melanie Carvalho and Graham Lowe and through ethnographic investigation she has co-curated several exhibitions including *An Archaeology of Race, Nurturing Ecologies* and *Describe a Landscape ...* . Her recent research monograph is entitled *Landscape, Race and Memory* (Ashgate, 2010). She is currently collaborating with artists on research entitled *An Archaeology of Race at the Museum*. Here she critically engages with the universalizing, imperial accounts of the theories of art and material culture from a postcolonial perspective.

Chapters

Caren Yglesias is a Professor in Landscape Architecture at the University of Maryland. She has expertise in landscape architecture design and practice. Her most recent monograph (2012) is entitled *The Complete House and Grounds: Learning from A.J. Downing's Domestic Architecture*. Caren is currently holder of an *American Association of University Women* (AAUW) fellowship (2012–13). She is also a significant contributor to contemporary dialogues on architectures for living via her blog: www.homesforliving.blogspot.com.

Claire Waterton is a Senior Lecturer in Sociology at Lancaster University, UK. Her research interests include: investigating the making and politics of scientific knowledge on nature and the environment; public perceptions of environmental issues and environmental risks; the relationship of scientific knowledge to contemporary environmental policymaking; science studies and the management of the rural environment.

Ian J. Winfield is a Researcher at the Centre for Ecology and Hydrology at Lancaster Environment Centre, UK. His research focuses on the management of freshwater fish undertaken within a Lake Ecosystem Group. This work is augmented by the use of state-of-the-art hydro acoustics to investigate fish abundance, distribution and size structure, together with aspects of their biotic and abiotic environments.

Ignaz Strebel is with the Centre for Research on Architecture, Society and the Built Environment at ETH Zürich (ETH Wohnforum – ETH CASE). He spent a number of years doing project work in geography and architecture at the Universities of Glasgow and Edinburgh (www.ace.ed.ac.uk/highrise/). His research touches upon science and technology studies of work place activities, decision-making and social complexity and related issues of urban change.

Jane M. Jacobs has a Chair in Cultural Geography at University of Edinburgh. Her research interests fall into two broad, and sometimes related, areas: postcolonial geographies and geographies of architecture. Her publications include: *Edge of Empire: Postcolonialism and the City* (Routledge, 1996), *Uncanny Australia: Sacredness and Identity in a Postcolonial Nation* (Melbourne University Press, 1998), and *Cities of Difference* (Guilford, 1998).

Judith Tsouvalis is a lecturer in Sociology at Lancaster University, UK. Her research interests include: the production of socio-natures, hybrids and cyborgs and how to do politics with things; the formation of knowledge-cultures and associated power-relations; symbolism, visualization and questions of representation of non-human actants in society; and the philosophy of time and space. Her present research is in Loweswater, in the English Lake District, with a focus on new forms of environmental governance.

Karen Wells is a lecturer in Geography at Birkbeck College, University of London. Her current research focus is on how international political economy including war, international law, and global capitalism impacts on children and childhood; visual research methods; and the visual cultures of childhood.

Mark Jackson is a Lecturer in Geographer at Bristol University, UK with interests in political economy and historical and cultural geography. His interests traverse the domains of postcolonialism, urban geography, social theory, political ecology and political economy, with a focus on theories of modernity, urbanity, political ontology, materialities of economy, science and technology and aesthetic representation.

Mike Crang is a Professor in Geography at Durham University, UK. He has written widely on issues of preservation, heritage, ruins and identity. Within this field he focused empirically upon on practices of public and oral history, photography and museums looking especially at examples in the UK and Sweden. From the angle of visual aesthetics and senses of temporality and rhythm, he has become interested in not just issues of preservation and conservation but also their converse – destruction, dereliction and decay – as a collaborator on the project *The Waste of the World*. His work on the material cultures of waste is attempting to rethink approaches to the commodity through emphasizing unbecoming things – that are both distasteful and unstable.

Mimi Sheller is Professor of Sociology and Director of the new Mobilities Research and Policy Centre at Drexel University, USA She also holds a continuing appointment as Senior Research Fellow in the Centre for Mobilities Research at Lancaster University (UK) and is founding co-editor of the international journal *Mobilities*. She is on the international editorial boards of the journals *Cultural Sociology*, and *African and Black Diaspora*. She has recently (2012) co-curated *LA Re.Play: An Exhibition of Mobile Media Art.*

Paul Frosh is Senior Lecturer at the Hebrew University, Jerusalem. His earlier research focused on photography and cultural production, especially the stock photography and visual content industries. More recently he has published work on consumer culture and national conflict, the construction of civil and national solidarity in Israeli television coverage of suicide bombings, 'media witnessing' (theorizing the relationship between contemporary media and practices of witnessing), and the aesthetics of pre-digital television. Paul is currently working on a book-length project about the relationship between mass media and imagination. He is also Chair of the Popular Communication Division of the International Communication Association (ICA).

Stephen Cairns has a Chair in Architectural and Urban Design at ESALA (Edinburgh School of Architecture and Landscape Architecture). His publications include *Drifting: Architecture and Migrancy* (Routledge, 2003) and *The SAGE Handbook of Architectural Theory* (2010), and he co-curated the Reciprocity exhibit at the 2009 International Architecture Biennale Rotterdam (See *Open City:Designing Coexistence*, SUN 2009).

Acknowledgements

This collection emerges from the dialogues and discussion held at the *Visuality/ Materiality: Reviewing Theory, Method and Practice* conference held at the Royal Institute for British Architects, in London in July 2009. We would like to thank our contributors and all conference participants. In particular we would like to thank Professor Elizabeth Edwards, Professor Paul Frosh and Professor Jane M. Jacobs for their plenary talks which inspired the subsequent discussions and dialogue.

We would like to give our warmest of thanks to Rosanna Raymond and performers from Te Kohanga Reo O Ranana and the community of Ngāti Rānana who co-convened the session *Being with Hinemihi*.

We are grateful to all of our colleagues that presented conference papers including: Catelijne Coopmans, Judith Tsouvalis, Claire Waterton, Jennifer Tomomitsu, Denise Amy Baxter, Sarah Wilson McKay, Julia Petrov, Tamami Fukuda, Gavin Perin, Linda Matthews, Inga Bryden, Deirdre Conlon, Deborah Gambs, Tim Dant, Elia Charidi, Matthew Tiessen, Neil Jenkins, Sondra Cuban, Jacquelyn Allen-Collinson, John Hockey, Sarah Teasley, Roberta Simpson, Caren Yglesias, Kathleen Coessens, Susan Salhany, Alexander Hope, Anaele Diala Iroh, Penny Tinkler, Suzannah Biernoff, Nikolaus Fogle, Eric Olund, Rebecca Coleman, Yok-Sum Wong, Derek Sayer, Patricia Ticineto Clough, Karen Engle, Petra Rethmann, Mimi Sheller, Mike Crang, Jessica Dubow, Colin Cremin, Julie Lord, Monica Degen, Anne Cronin, Sharon Lockyer, Diane Mavers, Nick Mahony, Barbara Adams, Adeola Enigbokan, Ryan Diduck, Craig Campbell, Kimberley Mair, Adam Kaasa, Mark Jackson, Annebella Pollen, Stephen Walker, Giovanni Porfido, Ian Heywood, John Potvin, S.H. Iradji Moeini, Teal Triggs, Claire McAndrew, Ana Araujo, Taisuke Edamura, Duan Wu, Katrina Brown, Rachel Dilley, Paola Jiron, Debbie Lisle, Jane Lydon, Courtney Baker, Anne-Marie Fortier, Karen Wells, Veronica Davidov, Nirmal Puwar, Linda Daley, Les Roberts, Paul Simpson, Jodi Polzin, Neil Walsh, Katie Lloyd Thomas, Eric Laurier, Fuyubi Nakamura, Ruth Panelli and Dean Sully. We would also like to thank film director Kuldip Puwar, for the debut screening of *Unravelling* at the conference.

The most important of acknowledgements goes to conference administrators Jan Smith (Open University) and Kathy Wood (Durham University), Chris Orton (Durham Design and Imaging Unit) and to the handful of postgraduate volunteers; without whose help the conference could not have run so smoothly.

Finally, but not least, our deepest gratitude goes to our editor Val Rose at Ashgate for all of her support and encouragement on this volume.

Also a personal thanks from Divya to Peter Kelly for his unerring support.

Chapter 1

Visuality/Materiality:
Introducing a Manifesto for Practice

Gillian Rose and Divya P. Tolia-Kelly

Amongst the various calls for theoretical re-orientations within the social sciences and humanities over the last twenty years, there has been an increased interest in the visual and material, evidenced by many new handbooks of visual culture (Heywood and Sandywell, 2011; Leewen and Jewett, 2001; Mirzeoff, 1998; 1999; Carson and Pajaczkowska, 2000), visual anthropology (Appadurai, 1988; Pink, 2005; 2006), methodology (Rose, 2001) and of material culture (Buchli, 2002; 2004; Hicks and Beaudry, 2010; Pink, 2004; Tilley, 2006), guides to visual methodologies and journals devoted to the visual or to the material. The aim of this contributed volume is to do something rarely done in these by-now substantial bodies of work, which is to attend to the relationships *between* the 'visual' and 'material', and to explore what kinds of new thinking might emerge in that intersection. The collection attempts to stage a respectful engagement with accounts of both the material and the visual, as they have emerged across a range of disciplines.

'Visual culture' (see Dovitskaya, 2005; Rogoff, 2001; Smith, 2005; 2008) has rapidly emerged as a privileged term for exploring 'the visual'. As a field of study, 'visual culture' responds to the myriad of shifts in visual media and its grammars. Included in the foundational lexicon of 'visual culture' are engagements with theories of Marxism, feminism, postcolonialism, identity and race, and 'visual culture' has a continuing relation to cultural studies in intellectual framing, texture and forms (compared with the Western imperial foundations of art history). This collection collates visual culture research that is reflexive about its theories, technologies and practices, and about its position as a realm of intellectual study that has experienced a set of refiguring, renewals and reincarnations such that cultural studies is no longer its only reference point. The politics of *visual culture* are enlivened further in this collection by including a focus on how researchers engage with *theory in practice*. In editing this volume we have privileged papers which delve into research as an attempt to account for the embodied politics present in the everyday material world. It is at this nexus that emergent visualities are enabling political revolutions, the 'war on terror' and fuelling the everyday geopolitical economies of cities, identities, histories, everydays and socialities. Through practical technologies there is a continuing mobilisation of communicative aesthetics which refigure our encounters with space, form, time, grammars of meaning and their habitual interpretation. The material turn for du Preez's (2008) account of art argues for a careful attendance to the stuff of art rather than a 'rhetorical deployment' of materiality (see Kearnes, 2003), which needs to be combined with a commitment to look (Bal, 2005).

The material 'turn', meanwhile, has been more about a *re*-turn; for some materialists at least, it is partly a response to a feared negation of materialities and those things that matter. Another part of this return has been a move towards restoring an alternative philosophical legacy to a denuded account of cultural materialism (Anderson and Wylie, 2009). These calls for a return embody a fear of an ephemeral account of culture and society, and a hope that 're-materialising' would reaffirm a formal politics of *materialim* (Jackson, 1989; Whatmore, 2006; Cook and Tolia-Kelly, 2010), rather than an *idealism* underpinning elements of the 'cultural' turn with its focus on language, text and poststructural accounts of postmodernism. These calls for materialism also respond to accounts of culture in society which were seen to be without connection to economy, society, situated bodies and the material infrastructures of societies' politics, inequalities and ideologies. Within the bounds of these material turns, however, the speculative, temporal, spatial and, above all, *visual* processes of becoming, enchantments and vibrancy (Bennett, 2001; 2010), hauntings (Stewart, 2007; Edensor, 2008) or indeed 'against' materiality (Ingold, 2007), and are all at the margins of what is seen to be core to the work of social science.

Neither turns have thus taken seriously the need for research on materiality that requires an understanding of the co-constitution of visuality and materiality. *Visuality/Materiality* emerges in this context as a collection which promotes the dialogues made possible in a space where these two modes of enquiry in their research are coconstructed. The scholarship represented here is reflecting research conducted in response to the call to *rematerialise*; but it also reflects 'more than representational' (Lorimer, 2005) research sites and practices where the cultures of the visual have exceeded the narrow, pedestrian promises of *matter* (see Coole and Frost, 2010; Barad, 2007).

Simultaneous with these theoretical debates, there has been a gravitational pull towards the visual, entailing a collective shift in praxis across the social sciences. In the realm of 'doing' research, text has become materially and visually framed, such that the contextalisation of visual forms as well as an urgent need to create the tools for the analysis of new media have become priorities. In the process of writing on art itself the text of academic practice has an embodied politics (see Hawkins, 2010). New modes of theorising the visual in anthropology (Pinney, 1998; 2011) and visual-historical anthropology (Edwards, 2001), as well as new visual elements of governance and security (Amoore, 2007), have enabled a congregation of political engagements and practice within the realms of visual culture. Reflected in this collection are these very creative cultures of thinking the 'visual' and 'material' which drive the scholars in this collection in the diverse realms of intellectual work in the disciplines of art history, anthropology, visual culture, geography, sociology, cultural studies, architecture and cultural geography.

In the contemporary research fields of visual culture and material culture there is a strong veer away from a 'pure culturalism' (Hicks, 2010, p 2). The senses, memory, body and history are part of the analytical process; as Buck-Morss (2002) argues, 'one needs all of one's senses to do justice to material reality' (p 328).

The aims of this collection are thus threefold: to theorise the interrelationship of materiality and visuality; to offer a series of empirical explorations of that interrelationship, which pay particular attention to research praxis; and to address questions of ethics in relation to difference, identity and power. The chapters were all presented at the *Visuality/Materiality* conference held in July 2009 in London. The *Visuality/Materiality* approach is evidenced here through research practices which are actively modest, contingent and partial, having at their heart political integrity and innovation. This is research as *practices* (and methodologies) which remember that the politics of *doing* the visual are as material as matter is visual and that both are engaged beyond the ocular.

Privileging Practice

This collection privileges how visual and material concerns are attended to in contemporary research through a focus on practice. Practice is what humans do with things. Some of the effects of some of those doings is to make things visible in specific ways, or not, and this approach thus draws attention to the co-constitution of humans subjectivities and the visual objects their practices create. This is somewhat different from enquiries based on looking, seeing, analysing and writing text; instead, it considers the (geo)politics of embodied, material encounter and engagement. This is a configuration of the practice of the visual and material in research that unravels, disturbs and connects with *processes, embodied practices* and *technologies*. Practice, processes and technologies are acknowledged as enabling intellectual enquiry to adhere to a path that is more-than theoretical, and more-than-representational (Lorimer, 2003), thus putting approaches to the *non-representational* beyond *pure theory*, and the processes of *representation* beyond *pure culturalism* as an analysis of the purely visual or material basis of text.

Here, we map this approach as an identifiable intellectual site that some researchers have been traversing and inhabiting for some time. The collection invites a recognition of this site of practice and process, which sits beyond disciplinary boundaries and their constraints. *Visuality/Materiality* is an emergent orientation of research practice that is inevitably critical and constantly reflexive of the power play between representation, text, practice and technologies of production, display and performance. The legacy of materialism within cultural theory is extended, enlivened and made meaningful through an approach that recognises a world of more-than signification through text, narrative, line and object. At the heart of the collection is an attentiveness to a reconceptualisation of the visual (through theory, method and practice), as an embodied, material, and often politically-charged realm. The critical argument at its heart is that the 'visual' and the 'material' should be understood as in continual dialogue and co-constitution. This co-constitution is also advocated and recognised here as being shaped through politics and in turn shapes politics at various scales. Thus there is no visual/material site of ideas, performance, phenomenon and practice which is secured away from the often violent, dirty, messy maters of surveillance, governance, money, rights and bodies. Yet what continues to happen, except usually in rather isolated pockets of anthropological research, is that visualities and materialities are considered separately. This collection argues that these fundamental approaches to cultural practices can be understood by prioritising the analytical context of human practices. What people do with the affordances of particular objects is, in part, to co-produce visualities.

The Cultural Logics of Sights and Things

The approach of the researchers in this collection is not concerned with applying the usual cultural logics in order to determine the meaning of objects and texts of visual culture. Instead of interpretation and meaning, treating visual and material as co-constituted has produced, for our contributors, a greater interest in matters of mediation, ethics, consumption, practice and translation. What has emerged as primary in the chapters are clusters of thinking and practice around the themes and questions of 'how things are made visible', 'which things are made visible' and, as a particularly explicit focus in three chapters, 'the politics of visible objects'. These are deliberately different from modes of thinking in visual culture that are about being critical, or having a 'good eye' (Rose, 2011), where the researcher looks at a text, separate from it, distinguishing it from others and being involved in a process of judgement (Frosh, this volume). Embeded in the *Visuality/Materiality* approach here is a concern with a situated eye, an attunement to the collective, multiple and embodied textures, sensibilities and productive meanings of the visual through the material, and vice versa. The focus is on questions of effect, histories, and ethics of engagement, interpretation, practice and process, which often fracture or displace the familiar fields of genre, media, audiencing and production. We can describe this as a concern with *ecologies of the visual*; where the co-constitution of visuality and materiality is in constant dynamic process and situated within networks, hierarchies and discourses of power.

Making things visible is just one of the effects of a practice approach to the co-constitution of visuality and materiality: of not thinking 'visuality' as simply observation, nor considering the 'material' purely as 'solid matter'. The question of *what* is made visible are critical to analysing using this approach. One example of a foundational model for visual analysis and one approach that has informed our expansion from visual materialism per se has been *iconography.* Iconography (in relationship to the cultural landscape) has been a framework of reading visual representations whilst privileging matters of politics and economy (Cosgrove, Daniels); it has sought to collapse the notion of representation as 'truth', but it is also a mode of inquiry which denaturalises the signification of aesthetics, grammars and icons within a frame. Lorimer (citing Wylie) terms iconography as a 'less deceitful veil to be pieced 'vertically' (in order to uncover power structures), than a complex texture to be searched horizontally' (della Dora *et al.*, 2011 p 4). The *Visuality/Materiality* approach avocated here is about claiming collective possibilities as well as embodied and phenomenological, whilst decentring the capturing, objectifying eye. This is where the visual is an embodied process of situation, positioning (Hall, 1990), re-memory (Morrison, 1990; Tolia-Kelly, 2004), encounter, cognition and interpretation. The *materiality* within our approach does not assume solidity of object and fixity of meaning (e.g. Dant, 1998), but incorporates the poetics of rhythms, forms, textures and the value of memory-matter engagement. Matter can be temporally and spatially unfixed, so that not only can the proverbial Proustian madeleines be evoked through the scent, touch (e.g. Brown *et al.*, 2011) sound and aesthetics of materials, but the sensory affordances of materials can also incorporate a pluralistic account of reactions and interpretations that link to *histories*, *memories* and *ecologies of seeing*, feeling and perceiving.

How Things are Made Visible

In this volume, Sheller, Jackson and Crang all locate the processes, practices and technologies that make certain things visible. These authors explore the naturalised *visibility* of materialities and tear away the seeming integrity of discourses, narratives and visualities which solidify our cultural logics of valuing and affixing meaning to their subjects. They differently subvert what is usually recognised, understood and seen within particular logics of modernity and enrich accounts of materials such as aluminium, ships and mass commodities. Sheller eloquently argues for understanding aluminium and its technologies through a 'visual semiotics for the technologically sublime'. Aluminium simultaneously is at the heart of capitalist 'dreams of de-materialisation' at the same time as being reliant on being mobilised for 'increased earthly destructions' (through military use) and resulting increased toxicity of land, peoples and oceans. Driving Sheller's account is a politics of materiality that is being shaped and economically consolidated in the late twentieth century in the forms and process of 'cybernetic economies running on semiotic superhighways' (Lash and Urry, 1994 cited in Sheller). The geopolitics of aluminium design, production, advertising and affective experience are unravelled by Sheller to expose mythologies through illustrating occluded histories, visualities and materialities of inequality, ecological degradation and neo-Imperial ecologies of seeing and governance.

Jackson's chapter takes Yiwu, China as a preeminent site where he unravels for us 'the architectures of the visible', in this city which is emblematic of China's nation building programme. Yiwu is a site of production of both aspirations of world and materials to assist us in the world, to practice the lifestyles of late capitalist modernity. Yiwu's contribution to aspirational politics are driven by the visualised futures of material living through the constellation of commodity opportunities and economy it innovates. Yiwu, in Jackson's account, is revealed as a critically important hub for firstly the production of the materialities of our global everyday consumption modes; the materialities of UNHCR, Disaster Aid, Wal-Mart, and Tesco. And secondly, Jackson (informed by Barad) illustrates how Yiwu 'the Commodity City' creates a material/visual commodity field through its conglomeration of city shops, exhibition centres, markets, outlets, advertising hoardings, economy, industry, and its ideological position in China's economy, 'an urban catalyst for aspirant narratives and their representation'. By tracing Yiwu's materialities from an atomic to a global scale we see how Yiwu inhabits a situation of being a 'contemporary cultural economy of consumption' which incorporates local grammars, and is powered and actively drives visualisations of ideologically driven notions of international consumption and identities that feature in our own locales visually and materially. Flowing, making mobile and empowering the transitions of ideological into material cities 'other' to Yiwu is the figure of the normally occluded shipping container, made visible in this account.

Crang starts by thinking fluidity in steel, and looks to philosophers Jane Bennett and Michael Serres for a departure point for thinking shipbuilding, steel and the materialities of living modernity through mobility rather than fixity; a vision of metal that is both material and creative. Ships are positioned at the heart of a notion of negative becoming; fluid, material, envisioned as torn from the usual discourses of bounded, fixed and solid to thinking about their 'breaking' as 'a negative becoming, or a sense of productivity that includes failure, disassembly and destruction'. The processes, practices of seeing these dynamic hulks of the sea is disturbed and refigured in Crang's account. The atomisation of metal is visualised as the source of becoming other both conceptually and materially here, presenting and making visible, a poetic account of routes of material that are not privileged in a notion of commodity, economies of the sea or indeed of steel. Crang pursues the question through photography and considers 'how does photography reveal, indeed revel, in the transcience it finds in this obdurate material through the grammar of the still image?' The matters of seeing and touching process through visual technologies and text fold into each other here to produce a *Visuality/Materiality* field that is more-than representational, but which is tuned into the economies of enchantment and profit.

Practices of Looking

Yglesias, Jacobs *et al.* and Tsouvalis *et al.* explore specific practices of *looking* at and with material objects, and explore the consequent seeing, envisioning and registering. Yglesias suggests training the visual imagination to enhance the material process of drawing, of communicating the mind's eye. Balancing the affective and pragmatic tools of practitioners is unravelled through her account of the role of creative and material practices, as they are intertwined in architectural work. Yglesias argues that 'seeing is more than an optical operation; understanding what is seen is a thoughtful activity', acknowledging this enables us to witness the process through which designers can 'create places of enriched experience' in two dimensional drawings. A particular attentiveness is needed, a phenomenological attunement to the practice of looking prior to drawing; feeling, experience and affective registers are thus translated onto the page. The aim for the author is to promote a notion of visual practice which engages with the matters of artistic intuition, situation, experience and the *doing* of drawing which results 'in a more truthful and sound manner' of technique.

Tsouvalis *et al.* also strive for a depth of seeing and a practice of seeing informed by multiple modes of seeing the English lake of Loweswater. Various practices of looking and seeing this site are presented in dialogue. Echo-sounding as a form of seeing through a hydro acoustic scientific survey is laid alongside the material visualisations and ethics of seeing that are promoted by locals and environmentalists. Rendering the algae of the lake as visible for the authors is an opportunity to do 'politics with things' (Latour, 2004). Based on the philosophies of material intra-action (Barad, 2007) the lake, and envisioning algae becomes a political setting with practical, ethical and political contingencies.

Jacobs *et al.* drive forward research on urban materialities (Latham and McCormack, 2004), by taking seriously the concept of assemblage and exposing the rich relationship between visualities and materialities. Their chapter sees the high-rise residential development Red Road not simply as an object of visualisation, but also as a technology that supports and depends upon what Cosgrove called 'vision in the sense of active seeing'. The sites of active seeing are the windows of Red Road. The window is a complex assemblage of things (material and immaterial): design specifications, material components (glass, jambs, frames), building and safety regulations, mechanisms such as hinges and locks, and decorative artefacts such as curtains. In unison with human users it becomes a purposeful aperture between the interior and exterior of a building: letting in light, ventilating, offering views (both out and in), as well as metaphysical opportunities. It also offers other, un-programmed opportunities: an opening to jump from or through which to throw rubbish. The chapter links the windows of Red Road and the viewing practices they afford and live alongside to visualities of other orders (avant garde architectural visions, urban visions, the visialisations of housing and building science, electronic surveillance, the glance).

The Ethics of Envisionings

Three further chapters focus in particular on examples of more intense power relations that can be articulated through specific enactments of visuality (see Poole, 1997) and materiality. Wells builds on previous work (2007) and presents a collage of practice which calls for a public recognition of what is seen, hidden allowed to be commemorated and the politics of public memory. Grammars of class, race, geopolitics of governance are all encompassed in her chapter which thickly describes the visceral outcomes of exclusion of certain visualities which become remainder, or marginal to others. The historical foundations of the body-politic of nation are at the heart of Puwar's account. Puwar takes the methodology of doing *visuality/materiality* further through the grammars of her own text. Expressing the limitations visually and materially of writing when dealing with the affective and body-politics of the everyday city, her writing exposes the positioning through which the city is revealed a counter-lens. Both Puwar and Wells urge us to witness, to reflect and *feel* the cityscape, so often images, represented and utilised in particular ideological accounts of history, memory and commemoration. Wells argues that seeing is a political practice of historical consciousness. Viewing, seeing and encountering sights are for Wells, require an embodied ethics of practice where a necessary logic of experiencing place requires an attunement to the sedimented history of a place. Everyday seeing, remembering and living is about an ethics of practice; the inclusion and occlusion of the histories of 'others' becomes then a collective responsibility for those traversing, seeing, viewing and situating themselves in place. Looking is a responsibility; a visceral, ethical and historically conscious practice.

Frosh (this volume) explores potential consequences of the mode of *inattentiveness*: how things that are visual are subject to routinised, inattention and distraction. This immediately challenges conventional accounts of interpretive approaches to the ethics of looking, seeing, and meanings texts, which assume a high degree of attention directed at images in order to look at them differently. Frosh comments on the ethics of a contemporary culture of the visual where seeing and looking are partial where a reciprocity between the transmission of images such as television and reception of the audience is not in synthesis. The social practices of inattention, where images become the background or 'wallpaper' to our everyday lives, becomes for Frosh a political problematic which he terms 'civic inattention'. The production of images of the material world is continually exposed to a lack of concern, care and subjugation to the 'eye'; no longer arresting, grasping or indeed holding or moving our attention. While this account of the collapsing of visuality through the practices of watching television risks rendering any political imperative to respond to 'war', 'terror', 'genocide', 'poverty' impotent, Frosh argues that it may also harbour the potential for a tolerant indifference towards difference.

Concluding Remarks

The nine chapters in this collection are all grounded in detailed empirical work, attentive to what is done in the world between people and objects. Those doings are of course extraordinarily diverse. From highly contested visualisations of the Caribbean to a visuality saturating the commodities in a Chinese exhibition hall;

from glances at and through a window to built and painted shrines and memorials; from the affordances of echo-sounded images to photographic images; from what might happen when an object is carefully drawn to what might happen when a television is inattentively watched; our contributors will take you to all these sites and the practices that take place there. They are all also highly attentive to the consequences of those practices. In those specific practices, what is made visible? (And what is rendered invisible?) How is it made visible, exactly – what technologies are used, and how, and what are the specific qualities of the visual objects thus enacted? And what are the effects of those visualised materialities and materialised visualities, particularly for the people caught up in those practices, as researchers, and as those researched? For while this introduction has offered a conceptual framework for approaching material and visual cultures, we would like to end our contribution to this collection by affirming the need for many more investigations of this kind: empirical, nuanced, alert. For it is only through such engagements with visual and material culture that we stand a chance of understanding just how contemporary culture is once again reshaping and reforming itself.

References

Amoore, L. (2007) Vigilant visualities: the watchful politics of the war on terror. *Security Dialogue*, 38, 215–232.

Anderson, B. and Wylie, J. (2009) On geography and materiality. *Environment and Planning A*, 41, 318–335.

Appadurai, A. (1988) *The Social Life of Things: Commodities in Cultural Perspective* (Cambridge: Cambridge University Press).

Bal, M. (2005) The commitment to look. *Journal of Visual Culture*, 2, 5–32.

Barad, K. (2007) *Meeting the Universe Halfway – Quantum Physics and the Entanglement of Matter and Meaning* (Durham and London: Duke University Press).

Bennett, J. (2010) *Vibrant Matters: A Political Ecology of Things* (Cambridge: Duke University Press).

Brown, G., Browne, K., Brown, M., Roelvink, G., Carnegie, M. and Anderson, B. (2011) Sedgewick's geographies: touching space. *Progress in Human Geography* (on-line version).

Buchli, V. (2004) *Material Culture: Critical Concepts in the Social Sciences*, Volume 1; Volume 3 (London: Routledge).

Buchli, V. (2002) *The Material Culture Reader* (London: Routledge).

Buck-Morss, S. (2002) Globalisation, cosmopolitanism, politics and citizen. *Journal of Visual Culture*, 1, 325–340.

Cook, I. and Tolia-Kelly, D. (2010) Material geographies. In: D. Hicks and B. Mary (eds) *The Oxford Handbook of Material Culture Studies* (Oxford: Oxford University Press).

Coole, D. and Frost, S. (eds) (2010) *New Materialisms: Ontology, Agency and Politics* (Durham: Duke University Press).

Dant, T. (1998) Playing with things: objects and subjects in windsurfing. *Journal of Material Culture*, 3, 77–95.

Della Dora, V., Lorimer, H. and Daniels, S. (2011) D. Cosgrove and S. Daniels (eds) (1988) The iconography of landscape: essays on the symbolic representation, design and use of past environments. *Progress in Human Geography* (on-line version March 2011).

Dovitskaya, M. (2005) *Visual Culture: The Study of Visual Culture after the Cultural Turn* (Cambridge: MIT Press).

Du Preez, A. (2008) (Im) Materiality: on the matter of art. *Image & Text*, 30–41.

Edensor, T. (2008) Mundane hauntings: commuting through the phantasmagoric working-class spaces of Manchester England. *Cultural Geographies*, 15, 313–333.

Edwards, E. (2001) *Raw Histories* (Oxford: Berg).

Hawkins, H. (2010) The argument of the eye? The cultural geographies of installation art. *Cultural Geographies*, 17, 321–338.

Heywood, I. and Sandywell, B. (2011) *The Handbook of Visual Culture* (Oxford: Berg).

Hicks, D. and Beaudry, M.C. (2010) *The Oxford Handbook of Material Culture Studies* (Oxford: Oxford University Press).

Ingold, T. (2007) Materials against materiality. *Archaeological Dialogues*, 14, 1–16.

Latham, A. and McCormack, D.P. (2004) Moving cities: rethinking the materialities of urban geographies. *Progress in Human Geography*, 28, 701–724.

Leewen, T.V. and Jewett, C. (2001) *Handbook of Visual Analysis* (London: Sage).

Lorimer, H. (2005) Cultural geography: the busyness of being 'more-than-representational'. *Progress in Human Geography*, 29, 83–94.

Mirzeoff, N. (ed.) (1998) *The Visual Culture Reader* (New York and London: Routledge).

Pink, S. (2004) *Home Truths: Gender, Domestic Objects and Everyday Life* (Oxford: Berg).

Pink, S. (2006) *The Future of Visual Anthropology: Engaging the Senses* (London: Routledge).

Pink, S. (2007) *Doing Visual Ethnography: Images, Media and Representation in Research* (London: Sage).

Pinney, C. (1998) *Camera Indica: The Social Life of Photographs* (Chicago: University of Chicago Press).

Pinney, C. (2011) *Photography and Anthropology* (London: Reaktion Books).

Poole, D. (1997) *Vision, Race, and Modernity: a Visual Economy of the Andean Image World* (Princeton: Princeton University Press).

Rogoff, I. (2001) *Terra Infirma: Geography's Visual Culture* (London and New York: Routledge).

Rose, G. (2001) *Visual Methodologies: An Introduction to the Interpretation of Visual Materials* (London: Sage).

Smith, M. (2005) Visual studies, or the ossification of thought. *Journal of Visual Culture Journal of Visual Culture*, 4, 237.

Smith, M. (2008) *Visual Culture Studies: Interviews with Key Thinkers* (London: Sage).

Stewart, K. (2007) *Ordinary Affects* (Durham: Duke University Press).

Sturken, M. and Cartwright, L. (2001) *Practices of Looking: An Introduction to Visual Culture* (Oxford and New York: Oxford University Press).

Tilley, C.Y. (2006) *Handbook of Material Culture* (London: Sage).

Tolia-Kelly, D.P. (2010) *Landscape, Race and Memory* (Farnham: Ashgate).

Wells, K. (2007) The material and visual cultures of cities. *Space and Culture*, 10, 136–144.

Whatmore, S. (2006) Materialist returns: practicing cultural geography in and for a more-than-human world. *Cultural Geographies*, 13, 600–609.

Chapter 2

Metallic Modernities in the Space Age: Visualizing the Caribbean, Materializing the Modern

Mimi Sheller

Introduction

Many technologies of mobility and speed depend on the special material qualities of aluminum, but they also depend on the *visions of mobility* that industry, artists, and advertisers put into motion: a visual semiotics for the technological sublime epitomized by the latest, fastest vehicles and streamlined objects that keep the world moving. This chapter offers a historical study of a specific material (aluminum) and its modes of visualization (in advertising, industry promotions, and public exhibitions), while linking this visual economy to a grounded political economy of its modes of production (bauxite mining) and consumption (including tourism). It shows how the design of a material culture of fast, light, streamlined modernism in the USA generated specific spatial geometries of seeing that were not merely representational, but were productive of the political subordination and economic appropriation of places conceived of as non-modern and peoples conceived of as racialized primitives. The visual imaginaries and material practices of a technologically advanced mobile modernity were co-constitutive of the "backwardness" of the Caribbean – a key location of both US bauxite mining and US tourism – not merely through the discursive continuation of colonial visual cultures, but via a more complex redeployment of colonial visual tropes as incitements to modern subjectivity within a modernizing material culture of consumer desire and touristic place-formation. Ultimately, though, people in the Caribbean have reconfigured such visualities and materialities, co-opting technologies of metallic modernity for counter-purposes of self-determination and the making of alternative modernities.

Light metal is what set the twentieth-century apart from past eras, and in many ways bequeaths to us today the distinctive look and feel of our "late modern" material culture based on aeriality, speed, and lightness. The strength and lightness of aluminum enhanced the mobilities of all modern transportation systems, first in the form of vehicles such as bikes, cars, trains, trucks, and above all airplanes; second in the aluminum-cabled electrical infrastructure and satellites that makes faster communication possible; and third in the soaring architecture of aluminum-skinned skyscrapers that symbolize modernity's vertical reach into the air along with the mundane objects of modern household life. Many of these technologies of light mobility depend on the special material qualities of aluminum, but they also depend on the *visions of mobility* that industry, artists, and advertisers put into motion: a visual semiotics for the technological sublime epitomized by the latest, fastest vehicles and streamlined objects that keep the world moving. The aluminum alloy airplanes and cars with aluminum engines, the sleek airport architecture and the glowing urban curtain-walls, the lightweight cans and convenient food packaging, the tubular alloy bikes and designer baby strollers that speed metropolitan people on their way, are the visible yet ignored infrastructure of mobile modernity.[1]

We get so used to the powers and possibilities of aluminum that we take it for granted and begin to forget it is there, tossing away billions of precious cans like so much trash, rather than the precious substance it is.[2] A multitude of metal products travels around the world, passing through our hands, lifted to our lips, lifting us off our feet. Light, fleet, it makes us feel super-human; we believe we can fly. Aluminum puts us in motion, leaving the earth's gravity. Meanwhile, far up in outer space a ring of aluminum satellites and space debris circles the Earth like a metallic halo, crowning our Space Age fantasies of mobile connectivity. Although it is often said that we live now in the information age, connected by the information superhighway, we still in fact depend on the vast movement of goods. Rapid, reliable, reasonably priced transport and communications systems are the *sine qua non* of our modern way of life, and it is largely aluminum that keeps the wheels turning and the virtual networks talking, texting, and twittering.

Aluminum makes the planes fly, the bombs explode, the rockets fire, and the satellites orbit. Out of the gravity of World War II bombs made from explosive powdered aluminum and the aluminum-fueled rockets of the Cold War, the aerial power of the Space Age dominated global politics. Heavy industry became lighter, ethereal electricity flowed through expanding circuits, and aluminum put the world on the path towards dreams of de-materialization, even as it enabled earthly destruction. Eventually the economies of the developed world became "economies of signs and space", as sociologists Scott Lash and John Urry described the late twentieth-century: cybernetic economies running on semiotic superhighways.[3] Our everyday lives remain, at the same time, furnished with goods inspired by the possibilities of aluminum design – from practical ladders and chic modernist chairs, to convenient cans and handy foil wrap, light laptop computers and sleek vehicles. We are bound up in metallic threads that fuse with our bodies, infiltrate our buildings, alter our way of life, and even make their silent way into our foods,

---

1   This chapter is based on my work in progress: Mimi Sheller, *Aluminum Dreams* (forthcoming).

2   For every six-pack of aluminum cans, energy equivalent to one can of oil is needed in their production.

3   Scott Lash and John Urry, *Economies of Signs and Space* (London: Sage, 1994).

makeup, antiperspirants, and medicines. Yet the ease of travel and the streamlined shining objects that serve today as the fetish objects of light modernity – bullet trains, space shuttles, Apple MacBooks – are grounded in the heavy (and dirty) industries of power generation, mining, refining, smelting, casting, and industrial fabrication. These remain part of a production process distributed around the world and controlled by a handful of huge multinational corporations – Alcoa, Rusal, Chinalco, Billiton-Rio Tinto.

These corporate powers have been accused of ignoring the rights of disempowered indigenous peoples in the tropics; consuming vast amounts of energy; contributing to global warming; and exploiting and polluting poorer countries.[4] The airy lightness of aluminum and its associated visualizations of metallic modernity were wrenched out of the tropical earth of specific places in the Caribbean (Suriname, Jamaica), in the remote aboriginal lands of Australia, and today in violently unstable places like Orissa, India and Guinea, West Africa. Just as anthropologist Sidney Mintz argued in *Sweetness and Power* (1986) that the modern Atlantic world was built upon sugar consumption in the age of slavery, we could say that aluminum offers a successor to that story: a late modernity built upon mobile consumption in the age of aluminum. The emergence of aluminum re-works the asymmetric material relations and visual circulations between urban centers of consumption and remote fringes of modernity where production takes place, while just as effectively erasing the modernity and humanity of the labourers who are just as much part of modernity's aluminum dreams. To tell the story of aluminum one must recombine the Northern world of mobility, speed, and flight, with the heavier, slower Tropical world of bauxite mining, racialized labour relations, and resource extraction. One also must acknowledge that the desired consumer objects and mobile practices of modernity cause ecological damage in the form of forest clearance, toxic red-mud, water and air pollution, and negative health effects on workers and nearby populations.[5]

In this chapter I want to consider how geometries of seeing and everyday embodied forms of moving, sensing, and visualization contribute to the performance of geopolitical (racial, national, gendered) identities and uneven splintered spaces.

---

4   Aluminum is produced using an electrolytic process in which a high current is passed through dissolved alumina in order to split the aluminum from its chemical bond with oxygen. To produce a ton of aluminum a typical smelter consumes at least 13,500 kwh (kilo-watt hours) of electricity, producing on average thirteen tons of carbon dioxide emissions per ton of aluminum, as well as the vast majority of the tetrafluromethane and hexafluoroethane emissions world-wide.

5   "Strip mining and [bauxite] ore processing produces about two and a half tons of wet mining wastes per ton of aluminum produced. It has historically led to severe soil erosion, as millions of tons of exposed earth and crushed rock were left to wash into streams and oceans. Strip mining destroys whatever wildlife habitat has existed above the mine, and is difficult – if not impossible – to re-establish even with intentional revegetation" (Jennifer Gitlitz, Trashed Cans: The Global Environmental Impacts of Aluminum Can Wasting in America, Container Recycling Institute, Arlington, VA, June 2002, p. 17).

By "geometries of seeing" I reference Doreen Massey's idea of "geometries of power" as a kind of spatial expression of power relations, but tie it more specifically to modes of visualization, materialization, and the systems of embodied meaning that they produce. Paying attention to practices of mobility and visualization alerts us to the simultaneous workings of visuality and materiality in producing subjects and spaces in constellations of differential power. In what specific circumstances does aluminum become visible or invisible, and for whom? What effects does the (in)visibility of aluminum have on spatialized and racialized relations of power? The chapter begins with a brief overview of the material culture of aluminum in the early to mid-twentieth-century, and its aspirational forms of visualization in the United States. It then turns to a close analysis of specific visual representations produced by the Aluminum Corporation of America (ALCOA) to promote tourism on Alcoa Steamships cruising through the Caribbean, where they also mined the bauxite that was the major source for the production of aluminum in the USA. As advertising and marketing incited consumers to value mobility, speed and lightness, their simultaneous materialization and visualization of tropical worlds of slowness, backwardness, and non-modernity were productive of the racialized geopolitical borders that still define the Caribbean as an "exotic" locale.

The emergence of aluminum-based practices of mobility, alongside modern ideologies and representations of that mobility, pivoted on the co-production of other regions of the world as backwards, slow, and relatively immobile – "bauxite-bearing" regions that would be mined by multinational corporations for the benefit of those who could make use of the "magical metal". Such relative mobilizations and demobilizations are constitutive of the connections and disconnections between North America and the Caribbean, with patents, tariffs, tax regimes, and military power locking in the spatial formations that allow discrepant or disjunctive modernities to exist side-by-side. Modern Caribbean underdevelopment, poverty, and political violence are as much a product of the Age of Aluminum as are gleaming iPhones or MacBook Air notebook computers with their "featherlight aluminum design" and promise of mobile connectivity at our fingertips. Finally, the chapter concludes by drawing out some lessons about the place of the Caribbean in the production of (American) modernity in the Space Age, as well as the production of "alterNative modernities" envisioned and materialized in the Caribbean itself.[6] What can we learn from Caribbean artists and theorists about more productive ways of visualizing and materializing parallel yet discrepant modernities? How can other meanings of technology and modernity be deployed against hegemonic visions?

---

6   Michel-Rolph Trouillot, *Global Transformations: Anthropology and the Modern World* (New York: Palgrave Macmillan, 2003).

Visualizing a Material Culture of Aluminum

If oil is the lifeblood coursing through the veins of the twentieth century, driving economies, politics, war, and diplomacy, then aluminum is the vessel that carries it. We usually recognize the centrality of oil to our modern world because of the wars that have been waged over it, or the consequences of global warming that have been linked to the carbon dioxide released when it is burned as fuel, but we seldom notice its quiet accomplice. The built environment that has grown up to support the oil economy is a material culture based in aluminum – especially in the high-power ACSR (Aluminum Cable Steel Reinforced) electricity lines that make up the power grid; the cars, trains, and airplanes with aluminum bodies that keep our economies flowing; and the aluminum-clad sky-scrapers that gleam in the centers of metropolitan power built on oil profits, starting with the aluminum spandrels and window frames of New York's Empire State Building (built at an astonishing speed of 25 weeks in 1930), and reaching an apogee in the aluminum curtain walls of the World Trade Towers that so spectacularly collapsed on September 11th, 2001, when struck by the aluminum airplanes turned into unexpected aerial weapons with devastating effect.

Aluminum played a crucial part in creating our contemporary world of mobility both in the material sense of enabling all of the new technologies that we associate with mobile modernity, and in the ideological sense of underwriting a world vision (and creative visualization) that privileges speed, lightness, and mobility.[7] The rise of an aluminum material culture occurred through a combination of new technologies, new aesthetics, and new practices of mobility. It was gradually recognized as an unusual material that was light yet strong, flexible and easily workable, quick to assemble and disassemble, easy to maintain, and reusable and recyclable. The weight of aluminum is about 1/3 of an equivalent volume of steel (with a specific gravity of 2.70 versus steel's 7.85), so engineers can use it to achieve dramatic weight reductions.[8] This has made it especially attractive in the aviation and transport industries, as well as in light-weight packaging and fasteners. From a pragmatic perspective it could contribute to weight savings, labor savings, cost savings and saved time in the building and transportation industries. But beyond these physical and practical properties it also became a

---

7   Tim Cresswell, in *On the Move: Mobility in the Modern Western World* (2006), describes mobility as consisting of physical movement, meanings and representations of movement, and embodied practices; all three dimensions are involved in my understanding of aluminum.

8   Aluminum's specific weight is 27 kN/m3, just one-third of the weight of copper or steel. Its melting point is approximately 660 degrees Celsius and with about "63 percent of the conductivity of copper, the most common conductor of electricity, for wires with the same dimensions", it can "replace copper in many applications" (Hugues Wilquin, *Aluminum Architecture: Construction and Details*, Basel, Berlin and Boston: Birkhäuser, 2001, pp. 12–13).

significant marker of modernity as a visual symbol of streamlined aerodynamic speed, clean gleaming lines, and the imaginative design of a futuristic world, as seen in early advertising campaigns by the leading American aluminum company ALCOA (see Figure 2.1).

Figure 2.1   Alcoa advertisement, 1930s

Aluminum put the world in motion and those new practices of mobility generated visual representations and symbolic economies revolving around the aesthetics of aerodynamic speed, accelerated mobility, and modernist technological futurism, including visualizations of cities and vehicles of the future. Thus, in addition to transforming the built environment and the infrastructures for mobility, the material culture of aluminum also influenced the ideas, beliefs, and meanings attached to movement in the twentieth-century. The industry's design and publicity departments played a crucial part in circulating these visual images and representations of mobility, instigating a wider culture of mobility and instilling a positive valuation of speed. As Florence Hachez-Leroy argues, the market for aluminum was initially non-existent, and promotion and publicity played a huge part in aesthetically enhancing the appeal of the curious new metal in the 1920s to 1930s, thus requiring the "invention of a market".[9]

The visions and possibilities of the light metal were expanded upon and elaborated by the inventors and "imagineers" of the future, who dreamed up new machines, buildings, and designs for living. For example, R. Buckminster Fuller, the eccentric inventor who designed the all-aluminum Dymaxion house in the 1930s, and the aluminum-bodied Dymaxion car in 1933,[10] went on to design the geodesic dome which achieved world fame through its exhibition as the centerpiece of the United States Information Agency's American National Exhibition in Moscow in 1959, and as the U.S. Pavilion for the Montreal Expo in 1967. The streamline esthetic was also advanced by designer Norman Bel Geddes, whose Motor Car Number 8 sported a teardrop shaped body. Geddes was the creator of the 1934 "Century of Progress" exhibition at the Chicago Worlds' Fair, and the General Motors Pavilion "Futurama" exhibit at the 1939 New York World's Fair (whose theme was "The World of Tomorrow"), which brought viewers riding on moving chairs equipped with sound into a futuristic 1960s world of fantastical skyscrapers, seven-lane highways, and raised walkways, "proposing an infinite network of superhighways and vast suburbs".[11] These future visualizations and simulations shaped the actual material cultures of later decades, as they were realized in material form by the industrial designers, architects, and engineers of modernity. Thus, Harvey Molotch argues that art "radiates visions that span across actors and industrial segments. Art brings enrollment", as inventors, investors,

---

9   Florence Hachez-Leroy, *L'Aluminum français: L'invention d'un marché*, 1911–1983, (Paris: CNRS Editions, avec le concours de l'Institut pout l'histoire de l'aluminum, 1999).

10   The only existing Dymaxion car was featured in an exhibition on Fuller's designs at the Whitney Museum of American Art, NY, in June–July 2008. See K. Michael Hays and Dana Miller (eds) *Buckminster Fuller: Starting With the Universe* (New York: Whitney Museum of American Art in Association with Yale University Press, New Haven and London, 2008).

11   Andrew Garn (ed.) *Exit to Tomorrow: World's Fair Architecture, Design, Fashion, 1933–2005* (New York: Universe Publishing, 2007), p. 62. And see David Nye, *American Technological Sublime* (Cambridge and London: MIT Press, 1994), Ch. 8.

power sources, patents, producers, promoters, and consumers are all enrolled into making and using something new in the world.[12]

This was not the first time a new material has transformed the world. Walter Benjamin wrote eloquently about the cultural impact of cast iron, which contributed new materials and visions for the transformation Paris between the 1820s and the1850s into a glittering city of arcades, grand boulevards, and exciting railways. In his *Arcades Project* he spins out a web of cultural connections from the cast iron structures of the gas-lit arcades, to the new department stores, and the cavernous railway stations, into an entire world of capitalist spectacle and new modern attitudes. Note that spectacle and visuality were central to this material transformation. This was the beginning of a movement towards modernity, and a fascination with the speed of the galloping stagecoach and soon the steaming locomotive thundering on its iron rails, with the instantaneity of daily newspapers and the fascination with kinetoscopes and moving pictures. Thus, the industrial technologies of material culture bear a close relationship to the visual cultures of each era, whether in the age of mechanical reproduction or digital communication.[13]

The discovery of aluminum smelting emerged from the 1850s to 1880s, the same period as the invention of cinema. Fast motion photography was built upon photochemistry and electric triggers, which made instantaneous photographs possible, just as chemistry and electricity together make aluminum possible.[14] As Rebecca Solnit has eloquently observed, it was the railroad baron Leland Stanford who supported the experimentations of photographer Eadweard Muybridge while he carried out his famous motion studies of race horses on Leland's Palo Alto estate, which later became Stanford University, the birthplace of Silicon Valley and the internet age. The railroad and the instantaneous photograph became the progenitors of mechanized moving pictures on perforated celluloid strips, as well as the violent means by which Native Americans were pushed off their lands (as the railroad expanded European settlement across the continent), only

---

12   Harvey Molotch, *Where Stuff Comes from: How Toasters, Toilets, Cars, Computers, and Many Other Things Come to be as They Are* (New York and London: Routledge, 2005, p. 87).

13   The combination of electricity and electrochemical production of metals has been called the second industrialization, replacing the canals, water-power, coal, iron, and steam power of the first industrial revolution. We might think of this as a shift from heavy to light modernity. Aluminum initiated this shift, and was later joined by other light substances such as plastics, fiberglass, and nylon, all spun off from the petrochemicals industry, and more recently silicon and titanium.

14   A technological revolution occurred with the invention of the electrolytic process for smelting aluminum in 1886. One of the great triumphs of modern chemistry, the dramatic race to invent and secure a patent on this electrolytic process for making aluminum was achieved simultaneously by a 23-year-old American, Charles Martin Hall, and a 23-year-old Frenchman, Paul Louis-Toussaint Héroult. See George David Smith, *From Monopoly to Competition: The Transformation of Alcoa, 1888–1986* (Cambridge and New York: Cambridge University Press, 1988).

to be preserved as shadow figures in the Wild West shows and the Hollywood westerns.[15] The parallel story of aluminum is also a national romance combining monopolistic industrial barons, new technologies of materialization (the making of metal objects) and new technologies of visualization (such as aerial vision and satellite vision), often exercised in colonial empires.[16] In the early twentieth-century aluminum began to open whole new vistas in the quest for speed, the new material culture of soaring urban heights, and the visual possibilities of illuminating the modern city with glass curtain walls and ubiquitous electricity. Aluminum would bring the apotheosis of speed and the new architectures of luminosity that nineteenth-century writers dreamed of, thrilled at, and feared. It eventually even enabled the flights to the moon first envisioned by Jules Verne, the science fiction writer who first recognized aluminum's potential in his 1865 novel *From the Earth to the Moon*, when this mysterious light element was still a rare and expensive precious metal.

The visuality/materiality of aluminum were an important aspect in the construction of what David Nye calls the "American technological sublime", which could be found in spectacular electrical displays, giant infrastructural projects, or impressive displays of machinic dynamism and speed.[17] In both the United States and Europe, aluminum quickly came to be used both in large-scale buildings and in small-scale everyday objects, indoors and outdoors, in public and in private settings. It embodied the new aesthetics of modernism and futurism, which in a way set the modern West apart from the rest of the world.[18] New modes of research, development, promotion and visual advertisement were designing *information* as much as objects, and marketing meanings and visions of the future as much as commodities. They were spreading the news of aluminum's possibilities, stoking dreams of its future potential, and assisting designers and fabricators in turning it into new products. And it is here that dreamers, designers, and artists played a crucial role in thinking an aluminum-inspired future into being. The light metal thus gave meaning and shape to the design of modernity in the twentieth-century and continues to be a significant signal of progressive values in twenty-first-century transportation, design, and architecture.

---

15   Rebecca Solnit, *River of Shadows: Eadweard Muybridge and the Technological Wild West* (New York: Penguin, 1993, p. 219).

16   On the development of aerial vision technologies and the role of aerial surveillance in the colonial world see Peter Adey, *Aerial Life: Spaces, Mobilities, Affects* (Oxford: Wiley-Blackwell).

17   David Nye, *American Technological Sublime* (Cambridge and London: MIT Press, 1994).

18   Aluminum played a crucial part in the vision of the Italian Futurists, and the industrialization of Mussolini's Italy and Hitler's Nazi Germany. Arnaldo Mussolini (brother of Benito), wrote in 1932: "We have often said: just like the nineteenth century was the century of iron, heavy metals, and carbon, so the twentieth century should be the century of light metals, electricity, and petroleum".

However, the industry's celebration of its own contributions to mobility, to technological advancement, and to global productivity actually mask the behind-the-scenes work that enabled it to lock in material immobilities (of technologies, capital, and people) grounded in global economic inequalities. Beyond the iconic material culture of American and European modernist mobility that aluminum companies envisioned and promoted, we must also consider the transnational *material contexts* inhabited by these multinational corporations. Through the use of patents, cartels, international trade regimes, anti-trust battles, negotiations with various states, and the benefits of military power, the industry monopolized control over the global movement of bauxite, electricity, and aluminum, interfering in the resource sovereignty of developing countries around the world. While the aluminum industry took off by promoting a gleaming aerodynamic modernism and super-mobility in its primary consumer markets in the United States, it simultaneously benefited from, and reproduced, a very different image of an as yet unachieved modernization in the "slow" Caribbean, and other tropical places where bauxite was found. Tourism and its modes of visual imagery played a crucial part in promoting this view of the Caribbean.

Thus, there was an implicit connection between the production of the material culture and visual image of modernity in the United States in the age of aluminum and the parallel consumption of raw materials and visual images of tropical backwardness in the Caribbean. More specifically, the design of a material culture (and *collective feeling*) of fast, light, streamlined modernism in the USA generated specific geometries of bi-focal seeing that were not merely representational, but were productive of the political subordination and economic appropriation of places conceived of as non-modern and peoples conceived of as racialized primitives. The visual imaginaries and material practices of a technologically advanced mobile modernity were co-constitutive of the "backwardness" of the Caribbean not merely as a discursive continuation of colonial visual cultures, but via a complex redeployment of colonial visual tropes as incitements to modern subjectivity within a modernizing material culture of consumer desire.

Crucially, though, physical movement, its representations and meanings as well as its lived practice, are each enmeshed in power relations and struggles. Various state actors, corporations, privileged groups, disprivileged groups, and challengers vie for control over mobilities and immobilities, deploying material culture and visual culture to stake their claims. So, states and peoples of the Caribbean region were closely involved in bargaining, negotiation, and political conflict over who would benefit from bauxite mining and from tourism, who would define Caribbean modernity, and how to challenge the hegemonic visualities and materialities of US imperialism.

Cruising the Caribbean: Visual and Material Circulations

Bauxite mining underpins a crucial connection between the production of the material culture and visual image of modernity in the United States and the parallel consumption of raw materials and visual images of tropical backwardness in the Caribbean. Surprisingly, the material circulation of mining and light metals is intimately linked with the visual circulation of images of the Caribbean, and the embodied practices of tourism that connect the Caribbean and the USA. In this section I offer a brief overview of a striking series of luxury magazine advertisements, promoting the Caribbean tourist cruises of the Alcoa Steamship Company from 1948 to 1958, the period in which the Aluminum Company of America became the biggest producer of aluminum in the world and depended to a large extent on bauxite mined in Jamaica and Suriname, the two largest exporters of bauxite in the world.

The Caribbean has long served as a site of tropical semi-modernity, set apart from the modern West through forms of colonial exploitation and imperialist exotification of its "colorful" people, "vivid" nature, and "dream-like" landscapes.[19] The cool Space Age futurism of aluminum modernity had to be constructed via its contrast with a backwards, slow world that happened to lie next-door: the American fascination with the steamy jungles of the tropics, the hybrid races of the Caribbean, and the image of the Third World as primitive, backward neighbors. This grammar of difference helped to construct what anthropologist Michel-Rolph Trouillot calls "the Savage Slot",[20] yet it depended on making *invisible* the power of US corporations to control and monopolize the mining and processing industries that make modern technology possible, and the military power that it enabled and required. It also made invisible the emerging modernities of the Caribbean itself, which were subsumed beneath the romantic naturalism of tourism.

The Alcoa Steamship Company played a special role in the Caribbean, not only shipping bauxite and some refined alumina to the USA, but also carrying cruise ship passengers, commissioning artists to depict Caribbean scenery, and even recording Caribbean music and sponsoring the Caribbean Arts Prize in the 1950s. The company operated three "modern, air-conditioned ships", each carrying 65 passengers, which departed every Saturday from New Orleans on a 16-day cruise, making stops in Jamaica, Trinidad, Venezuela, Curaçao, and the Dominican Republic. They also ran several freighters out of New York, Montreal, and New Orleans, which carried 12 passengers and made longer, slower trips; these ships delivered bauxite and alumina for Reynolds and Kaiser (after a major anti-trust

---

19   Mimi Sheller, "Iconic Islands", in *Consuming the Caribbean* (London and New York: Routledge, 2003); Krista Thompson, *An Eye for the Tropics* (Durham: Duke University Press, 2006).

20   Michel-Rolph Trouillot, *Global Transformations: Anthropology and the Modern World* (New York: Palgrave Macmillan, 2003).

law suit opened up the industry to competitors) as well as for the dominant global company Alcoa.

The same economic, political, and spatial arrangements that locked in huge market advantages for transnational mining corporations simultaneously opened up Jamaica for tourism mobilities. Tourism then instigated the circulation of new visual representations of movement through the Caribbean. While Alcoa promoted novel products, modern skyscrapers and metallurgical research and development in the USA, the company envisioned the Caribbean as a source of bauxite, a potential market for "superior" American products, and a timeless destination for tourists traveling on its modern ships to safely step back into the colorful history, exotic flora, and quaint folkways of diverse Caribbean destinations. While these modes of imaging are not surprising (and continue in other ways today), what is striking is the degree to which they diverge from the futuristic images of super-modernity that were simultaneously being promulgated in the US consumer market for aluminum products. Appearing in the luxury publication *Holiday* magazine, the images seamlessly meld together tourism, business travel, bauxite shipment, and cultural consumption, yet carefully detach these Caribbean mobilities from the futuristic super-modernity being envisioned and promoted at home in the USA in magazines such as *Fortune*. They also strikingly ignore or erase the presence of modern technology in the Caribbean, including US military bases, and the emerging infrastructure of airports, modern ports, and eventually rocket-launching facilities, as discussed below.

The first series of advertisements, which ran from 1948 to 1949, was created and signed by the graphic artist Boris Artzybasheff. A Russian émigré to the United States, Artzybasheff was most well known for his cover portraits for *Time* magazine, his colorful series of Shell Oil ads, and his surreal drawings of anthropomorphic machines. For the Alcoa Steamship Company he produced an unusual series of portraits of Caribbean people, each surrounded by a gorgeous tapestry of flowers, fruiting plants, and native fauna, creating a story-book impression of diverse exotic races, colorful landscapes, and romantic adventure. A Carib mother and child, for example, are depicted as timeless primitives, holding up fruit and peering out from dark eyes and exotically painted faces (see Figure 2.2). Although the images appear at first to promote cultural encounter and ethnological curiosity within touristic contact zones, such typifying images circulate within a long lineage of tropicalizing representations of Caribbean islands and people.[21]

---

21   W.J.T. Mitchell, *Landscape and Power*, second edition (Chicago: University of Chicago Press, 2002); Richard Grove, *Green Imperialism: Colonial Expansion, Tropical Island Edens, and the Origins of Environmentalism, 1600–1860* (Cambridge: Cambridge University Press, 1995); Candace Slater, "Amazonia as Edenic Narrative" in William Cronon (ed.), *Uncommon Ground: Rethinking the Human Place in Nature* (W.W. Norton, 1996); Sheller, *Consuming the Caribbean*; David Arnold, *The Tropics and the Traveling Gaze: India, Landscape and Science, 1800–1856* (University of Washington Press, 2006).

They serve to set apart the modern US consumer/tourist from the ersatz primitive populations of the tropics.

The text of the Artzybasheff series emphasizes the swashbuckling colonial history of the Caribbean, which marked the region with diversity: "if you look carefully you'll see how the distinctive architecture, languages and races of this area have been blended by centuries into interesting new patterns". Each image

Figure 2.2   Boris Artzybasheff, Alcoa Steamship Co. advertisement, *Holiday* **magazine, 1948**

represents an example of racial blending or distinctiveness, portraying a male or female racialized persona in typical costume, including Carib Indians, various types of Afro-Caribbean or racial "blends", East Indian Hindu and Muslim, and even various types of Creole "whites" of Spanish, Dutch, or Anglo-Caribbean origin. This kind of typifying imagery relates to earlier colonial racial typologies and Spanish *castas* paintings, which attempted to portray all of the racial "types" found through mixtures of different kinds. Each image includes a distinctive flower, foliage, and often a typical bird or butterfly, suggesting a kind of natural history that conjoins island people and wildlife, naturalizing races through attachment to specific places. This is a mode of visualization that solidifies Caribbean "difference" as a material, natural, textural distinctiveness that sets "the islands" apart from the mainland world inhabited by allegedly modern subjects. The images paper over the ethnic, class, and color hierarchies that fanned political unrest throughout the post-War Caribbean; they offer not only a visually flattened perspective, but also a historically and socially flattened one. Above all, they produce a visual grammar of difference.

The protean Caribbean appears here as a series of renaturalized yet traditional places, outside of modernity yet accessible to the mobilized tourist. Special editions of some of the images, suitable for framing, could also be ordered by mail, creating a Caribbean souvenir to take home. Thus, the adverts and the images circulate in a visual economy of touristic consumption, but one that is specifically enabled by the modern light mobility of the American consumer who lives in a world of fast modernity enabled by the very aluminum that drives the Alcoa ships to transport bauxite from the Caribbean, and carry tourists in. As the industry incites the design and consumption of new modern products in the US consumer market, it simultaneously produces the modern subject who will consume them through a contrast with the slow, backward, romantic tropics. The ships themselves were built with extensive aluminum fixtures and fittings, materializing the possibilities of advanced technology and modernist design.

While suggestive of the complex global mobilities of people and cultures, these historically anchored images fix the Caribbean in time as a series of romantic remnants that exist for the edification and consumption of the modern, mobile traveler. Only a Venezuelan worker placed in front of what is described as a "forest of picturesque oil wells", holding a sturdy wrench amongst some delicate pink flowers, hints at a modern industrial economy taking shape in the Caribbean, but one that is described as "*hungry* for American-made products – and all that their superiority represents". The florid feminization of the worker and naturalization of the industrial landscape suggest attempts to fit industrialization into older tropes of Caribbean island paradise. Thus, even as US businesses and military bases transform the material culture of the region, the image of the Caribbean as exotic, backwards, and romantic is preserved.

The second series of Alcoa Steamship adverts, published in 1951–52, depicts botanical paintings of tropical flowering trees by the respected botanical illustrators Harriet and Bernard Pertchik. These images tap into a long tradition of botanical collection and illustration of tropical plants by colonial naturalists, who collected material in the Caribbean and incorporated it into systems of plant classification and medical knowledge.[22] Botanical collection is linked to forms of colonial control over nature and an imperial gaze that merges the picturesque with the pragmatic use of natural resources and uses both as a justification for imperialist appropriation of land and resources for their alleged "improvement"; this prospecting imperial gaze was later reinvented as a key constituent of "the tourist gaze".[23] These adverts reiterate a kind of visual nostalgia for colonial botanizing and natural history, while including in their lower corner small sketches of white tourists relaxing on board the cruise ships, suggestive of the contrasting modernity and luxuries of the leisure class who can enjoy a cruise through the tamed islands of the Caribbean. It is American tourists who move, while the islands and Caribbean coastal countries are collected, catalogued, and made known as sites of natural beauty, even as their interior forests were in some cases being cleared and strip mined for the precious red bauxite ore below them.

The third Alcoa Steamship Co. series, which ran from 1954 to 1955, is a striking set of folkloric portrayals of musical performances, parades, or dances, both religious and secular, by the graphic illustrator James R. Bingham (see Figure 2.3). Readers are encouraged to write in to purchase 45 rpm recordings of the music that accompanies some of the dances, including the sensationalized Banda dance of Haiti, the Joropo of Venezuela, the Merengue of the Dominican Republic, and the Beguine of the French West Indies.[24] Other adverts represent the Pajaro Guarandol "folk dance of the Venezuelan Indians", the steel pan and "stick dance" of Trinidad, the "Simadon" harvest festival of Curacao, the folk dance of the Jibaros of Puerto Rico, and the John Canoe dancers of Jamaica whose costumes date back to the eighteenth-century and possibly to West Africa. This series again connects touristic consumption of musical performances from across the Caribbean with an almost ethnographic project of investigation of traditional cultures and people who persist outside of modernity, a remnant of the past available for modern cruise tourists to visit.

Recent studies in visual culture show how painted and photographic images of the Caribbean constantly re-inscribe picturesque island landscapes, romantic tropical nature, and hybrid mixtures of exotic races. Such images have frequently been used to market the region to both business investors and tourists. They have also had an impact on re-shaping the built environment and natural landscape

---

22   David P. Miller, "Joseph Banks, Empire, and 'Centers of Calculation' in Late Hanoverian London" in *Visions of Empire: Voyages, Botany, and Representations of Nature* (Cambridge, Cambridge University press, 1996).

23   John Urry, *The Tourist Gaze*, second edition (London: Sage, 2002); Arnold, *The Tropics and the Traveling Gaze*; Sheller, "Iconic Islands", *Consuming the Caribbean*.

24   Also crucial here are sonic circulations of recorded music, performers, and dances – a topic I cannot fully address here, except to suggest that we must also pay attention to the intersections of visuality and materiality with auditory and other sensuous geographies.

Figure 2.3   James R. Bingham, Alcoa Steamship Co. advertisement, *Holiday* magazine, 1954–55

to fit external expectations and serve the needs of both local elites and foreign corporations. As Krista Thompson has shown in her study *An Eye for the Tropics*, photographic images of the Caribbean produced for tourist markets in the early twentieth-century tropicalized nature by emphasizing lush and unusual plants, exoticized local people by showing them in rustic and primitive settings, and erased signs of modernity such as electric power lines or newer urban areas. Typical imagery such as the palm-fringed beach played on the symbolic resonances of the palm tree to materialize these islands as places of leisure and relaxation, rather than toil and poverty. And folkloric music was played by costumed performers for the benefit of tourists.

National elites had an interest in furthering these projects of self-exotification for the tourist market, just as they had an interest in encouraging foreign investment, whether in agriculture or mining. They tried to keep their towns looking quaint and not too modern in order to encourage tourism.[25] When modern buildings such as a new Hilton Hotel were built in the 1950s or 60s, they became enclaves of modernity for visiting tourists from which the local populace was excluded except as service workers. Yet seldom has the direct connection between the two industries been noted: the mining of bauxite made possible the mobilities of tourism, while the

---

25   Thompson, *An Eye for the Tropics*. It was often the same corporations involved in commodity trading and tourism: ships laden with bananas or bauxite on the way north carried passengers on the return journey, while businessmen interested in investing in the region stayed at hotels owned by the multinationals, and took their families on cruises.

touristic visualization of the Caribbean supported the materialities of dependent development that kept the Caribbean "backwards" and hence quaint to visit.

So in a sense the *absence* of aluminum architecture, vehicles, power lines, and designer objects came to define Caribbean material culture, which in contrast came to be associated with the rustic, quaint, vernacular, hand-made island tradition, using natural materials and simple processes. However, the people of the Caribbean were at the same time contesting such images, insisting on their own modernity. Independence movements in the post-War period began to call for self-rule, migrants to London, New York, and other metropoles, along with the radio, carried styles of modern urban cultural consumption back to the Caribbean, and Caribbean styles of modernity were themselves carried into the metropole, so-called colonization in reverse. Writers, musicians, intellectuals, and artists grappled with the meanings of Caribbean modernity, and produced their own visualizations of the Caribbean past, present, and future. In the final section I turn to the arrival of the Space Age in the Caribbean, the state practices that have contested its materialities, and the artistic practices that have contested its visualities.

Modern Materialities, Caribbean Visualities

In October 1953 the British government, with US support, forcibly suspended the constitution of Guyana and deposed the labor-left government of Cheddi Jagan, elected by a majority under universal adult suffrage, which had threatened to take back mineral resources and move the colony towards independence.[26] In Jamaica, major renegotiation of the terms of bauxite royalty payments and taxes was undertaken by People's National Party Chief Minister Norman Washington Manley (one of the founding fathers of Jamaican independence) in 1956–7, based on the principle that "countries in the early stages of economic development ought to derive the largest possible benefits from their natural resources. They ought not to be regarded merely as sources of cheap raw materials for metropolitan enterprises".[27] Following tough negotiations, the 1957 agreement re-set the royalty which led to a substantial increase in revenues to the Jamaican government, contributing more than 45 per cent of the country's export earnings by 1959.[28]

In October 1957 the Soviet Union launched the first Sputnik satellite, a small aluminum orb that triggered the Space Race with the United States. This, along with the Korean War, made aluminum an even more crucial "strategic material", and the US government began stockpiling it. Jamaica moved from supplying

---

26 O. Nigel Bolland, *The Politics of Labour in the British Caribbean* (Kingston: Ian Randle and London: James Currey, 2001, p. 617–21).

27 Norman Washington Manley to Kaiser Bauxite Company, 23rd May 1956, cited in Carlton E. Davis, *Jamaica in the World Aluminum Industry, 1838–1973*, Vol. I. (Kingston and London: Jamaica Bauxite Institute, 1989, p. 189–90).

28 Davis, *Jamaica*, p. 229, 251.

about one quarter of all US bauxite imports in 1953 to over one half in 1959, with 40 per cent of total shipments of crude bauxite and alumina between 1956 and 1959 going into the US government stockpile.[29] Jamaica achieved independence in 1962 and by the 1960s "was the world's largest producer of bauxite" according to historical sociologists Evelyn Huber Stephens and John Stephens.

> In 1965, the country supplied 28 percent of the bauxite used in the market economies of the world … [and] bauxite along with tourism fueled post-war Jamaican development and the two provided the country with most of her gross foreign exchange earnings.[30]

In 1962, coincidentally, not only did Jamaica achieve independence, but the first two Americans to orbit the earth, John Glenn in the craft Friendship 7 and Scott Carpenter in Aurora 7, happened to splash down in their aluminum capsules in the Atlantic Ocean close to the nearby Turks and Caicos Islands, and were brought to the US Air Force Base on Grand Turk for debriefing and meetings with US Vice President Lyndon B. Johnson. One imagines the top-secret space-craft washing up on a remote coral cay like something out of a James Bond novel; after all, Ian Fleming wrote twelve of the Bond stories at Goldeneye, his plantation retreat in the hills above Kingston, and his hero was named after the respected ornithologist who had authored the first authoritative field guide to *Birds of the West Indies*, from which Artzybasheff may have copied some of his birds.[31] Amidst the dying embers of the British Empire, Fleming conjured up the juxtaposition between super-modern Agent 007's extreme forms of hyper-mobility and the sinister tropical profusion, ancient Voodoo rituals, and racial/sexual intrigue of the "primitive" Caribbean. The fast metallic contours of Bond's line of Aston Martin cars, beginning with the 1965 model and up to the most recent DBS and DB9, built around a very high-tech "bonded aluminum Vertical Horizontal structure", embody his technological prowess, and the technological sublime spectacle of speed.[32] All the more so set against the seemingly "primitive" Caribbean backdrop. But the Caribbean was already slipping from British grasp.

Also in that same year of Jamaica's independence, the 50-foot antenna of the American Telstar satellite, which used 80,000 pounds of aluminum, beamed the first satellite television pictures back to a transmission station in Maine, ushering

----

29   Carlton E. Davis, *Jamaica in the World Aluminum Industry, 1838–1973*, Vol. I. (Kingston and London: Jamaica Bauxite Institute, 1989, p. 251).

30   Evelyn Huber Stephens and John Stephens, *Democratic Socialism in Jamaica: The Political Movement and Social Transformation in Dependent Capitalism* (Princeton: Princeton University Press, 1986, p. 26).

31   James Bond, *Field Guide to Birds of the West Indies: A Guide to all the Species of Birds Known from the Greater Antilles, Lesser Antilles and Bahama Islands* (The Macmillan Company, 1947).

32   <http://www.astonmartin.com/eng/thecars/dbs/control accessed 22/12/2010>.

in the dawning of satellite telecommunications. Seven years later aluminum made possible the first human landing on the moon in 1969, when Neil Armstrong and Buzz Aldrin stepped out of their landing craft onto the Sea of Tranquility while the mother ship *Columbia* orbited the moon. Around the same time, the New World Group of economists formed at the University of the West Indies and began to publish scathing critiques of foreign capital and the economic underdevelopment of Jamaica and began to call for the nationalization of the Jamaican bauxite industry in the early 1970s. In 1973 Prime Minister Michael Manley's People's National Party government, which was trying to implement democratic socialism in Jamaica,

> opened negotiations with the aluminum TNCs on acquisition of 51 percent equity in their bauxite mining operations, [...] acquisition of all the land owned by the companies in order to gain control over the bauxite reserves, and a bauxite levy tied to the price of aluminum ingot on the U.S. market.[33]

In March 1974, inspired by the success of OPEC, a bauxite producer's cartel known as the International Bauxite Association (IBA) was set up and was quickly able to double the price of bauxite on world markets.

* * *

The Caribbean may appear as an afterthought in the Space Age, as imagined in the United States; yet it seems fitting that in March 2008 the first European Space Agency Automated Transfer Vehicle, which docked successfully with the International Space Station, was launched on an aluminum Ariane 5 rocket from Kourou, French Guiana. The vehicle is appropriately named "Jules Verne". Gazing towards the heavens, an observer of the historic launch might not have noticed the displaced Saramaka Maroons living as non-national migrants on the fringes of the former French penal colony of Cayenne. These descendants of runaway slaves who gained treaty rights to an independent territory in the jungle interior of neighboring Dutch Guiana in the eighteenth-century lost a huge portion of their homeland in 1966–67 when Alcoa built the Afobaka hydroelectric dam and artificial lake to power their aluminum smelter at Paranam, displacing thousands of Maroon villagers.[34] These curious Caribbean footnotes to Modernity's outer space achievements ought to draw our attention back down to the ground of Suriname, Guyana, and Jamaica, where an analysis of the mobile *and* immobile material cultures afforded by aluminum can elucidate not only the cultural history

---

33 Evelyne Huber Stephens, "Minerals Strategies and Development: International Political Economy, State, Class, and the Role of Bauxite/Aluminum and Copper Industries in Jamaica and Peru", *Studies in Comparative International Development*, Vol. 22, No. 3 (Fall 1987, pp. 63–4).

34 Richard Price, *Travels With Tooy: History, Memory and the African American Imagination* (Chicago: University of Chicago Press, 2007).

of technology, design, and popular culture, but also broader currents of global relationality and relative positioning.

A fascinating interrogation of technology, modernity, and Caribbean materiality can be found in the recent work of Bahamian artist Tavares Strachan, with which I conclude.[35] Strachan's "Orthostatic Tolerance", is a series of inter-related installations documenting the activities of the Bahamian Aerospace and Sea Exploration Center (BASEC), an ersatz space agency created by the artist and modeled on NASA. Aspects of the project were shown at the Institute of Contemporary Art in Philadelphia in 2009; at Grand Arts in Kansas City, MO, in February to May 2010, with the title "The Orthostatic Tolerance: Launching into an Infinite Space"; and at MIT's List Visual Arts Center from May to July 2010, with the title "Orthostatic Tolerance: It Might Not Be Such a Bad Idea if I Never Went Home".[36] Strachan's show at the ICA in Philadelphia documented the BASEC launch of a sugar-fueled glass rocket over a Bahamian reef. While raising a smile with its mock insignia, cosmonaut training and video of a beach-side launch that ends in a broken rocket, this project nevertheless addresses poignant themes of the Caribbean's relation to modernity, science, technology and art. The artist himself appears at the rocket launch clad in a full-body white bio-hazard suit, as if prepared to encounter alien worlds. The rocket's glass is made from island sand, and the fuel from island-grown sugar cane, representing the transformation of "basic" Caribbean natural substances into "advanced" scientific technologies. The alchemical processing of sugarcane into fuel references the history of plantation slavery in the New World, while pointing towards the powerful energy in cane sugar that today promises the biofuels of the future. The loss of the glass rocket in the sea echoes other projects in which Strachan submerges blown-glass objects in tanks of mineral oil, so they are barely visible. The BASEC project thus resonates with the obscured visibility of the Caribbean presence in outer space exploration.

From his Caribbean perspective on sea, sky, and space Strachan's work indexes the long history of planetary exploration that has its origins in the European "voyages of discovery" to the New World, where Christopher Columbus first sighted land in the Bahamas in October of 1492. The Grand Arts project shown in Kansas City includes video of cosmonaut training that Strachan undertook at the Yuri Gagarin Cosmonaut Training Center in Star City, Russia. This involved a space capsule half-submerged in an indoor swimming pool, and video footage of the artist donning a space suit and entering the water like a scuba diver in practice

---

35   Born in 1979 in Nassau, Bahamas, Tavares Strachan is based in New York, NY. Strachan holds a BFA from the Rhode Island School of Design and an MFA from Yale University. Strachan's work has been written about in *The Brooklyn Rail*, *New York Magazine*, *The New York Times*, *Artforum*, *Art Papers*, *Flash Art*, and *Art in America*.

36   "The Orthostatic Tolerance: Launching into an Infinite Space", 2010 Grand Arts, Kansas City, MO, 5 February through 3 April, 2010; "Orthostatic Tolerance: It Might Not Be Such a Bad Idea if I Never Went Home", 2010 MIT List Visual Arts Center, 7 May through 11 July 2010.

(see Figure 2.4). He is plucked dripping out of the water, much like Gus Grissom being rescued from the Bahamian seas in 1961, as portrayed in the film "The Right Stuff" (dir. Philip Kaufman, 1983), or like John Glenn and Scott Carpenter in their 1962 splash-landing off the Turks and Caicos. Indeed, in 1961 US astronaut Virgil "Gus" Grissom infamously ended his 15-minute suborbital flight with the accidental sinking of his capsule Liberty Bell 7 in the Atlantic Ocean off of Grand Bahama Island (where it was finally recovered in 1999 by a Discovery Channel expedition). Works of Caribbean space exploration tap into this simultaneously irreverent yet serious intersection of popular culture, high technology, and Cold War politics.[37]

Figure 2.4   Tavares Strachan, still from "Orthostatic Tolerance: It Might Not Be Such a Bad Idea if I Never Went Home"

37   See also Gerald Horne, *Cold War in a Hot Zone: The United States Confronts Labor and Independence Struggles in the British West Indies* (Philadelphia: Temple University Press, 2007).

In facing towards an exploratory Caribbean future with "basic" tools, and even naming BASEC's lunar surface rover after the first African-American astronaut, Robert Lawrence, who died in a launch explosion, Strachan optimistically performs what Guyanese poet, novelist, essayist, and dramatist Wilson Harris describes as "breaking fixed linear ruling patterns into non-linear simultaneous movement of such patterns forwards and backwards. Such simultaneity brings us into the mystery of timelessness and helps the past to be re-creatively potent".[38] Harris himself does this through referencing alchemy, native cosmogonies, quantum physics, hypotheses on time and space, dreams, and aboriginal cultures and religions, all elements likewise to be found in Strachan's work, even if subconsciously deep inside its genesis. Moving us forward and backward in time, this analytical lens suggests that the metallic modernity of aluminum/bauxite might also be made visible in the Caribbean.

Strachan's work arguably takes part in an "often ignored strain of New World African culture: a techno-visionary tradition that looks as much toward science-fiction futurism as toward magical African roots". Similar strains are found in the work of poet and theorist Kamau Brathwaite, for example. Cultural Critic Mark Dery further suggests that,

> African-American culture is Afrofuturist at its heart, literalizing [William] Gibson's cyberpunk axiom, 'The street finds its own uses for things.' With trickster elan, it retrofits, refunctions, and willfully misuses the technocommodities and science fictions generated by a dominant culture that has always been not only white but a wielder, as well, of instrumental technologies.[39]

We can conclude with one final image, of a Jamaican recycler, melting down scrap aluminum in a home-made smelter (see Figure 2.5).[40] From this backyard operation, the detritus of modernity is up-cycled into newly crafted products – souvenir plates bearing the likeness of Bob Marley, light enough for tourists to carry home in their suitcases, valuable enough for some inventive street entrepreneurs to make a living. So Jamaica survives, reinventing modernity and projecting its own visualities and materialities of the future.

---

38   Wilson Harris, "The Mystery of Timelessness", in Kathleen Gyssels and Bénédicte Ledent (eds) *The Caribbean Writer as Warrior of the Imaginary* (Rodopi, 2009, p. 26).

39   Mark Dery, "Black to the Future" accessed at <http://www.detritus.net/contact/rumori/200211/0319.html>. See also Ken McLeod "Space oddities: aliens, futurism and meaning in popular music", *Popular Music* (2003), 22: 3: 337–355. In his iconic cyberpunk novel *Neuromancer*, William Gibson envisions a self-sufficient Rastafarian space colony called Zion, also hinting at an alterNative Caribbean presence in outer space.

40   See David R. Dodman, "Making a Living: Small-Scale Metal Smelting as a Livelihood Strategy in Kingston, Jamaica", in Rivke Jaffe (ed.) *The Caribbean City* (Kingston: Ian Randle, 2008, pp. 227–46).

Figure 2.5   Jamaican recycler

Source: Photo credit and permission: Rivke Jaffe

References

Arnold, D. (2006) *The Tropics and the Traveling Gaze: India, Landscape and Science, 1800–1856* (Seattle: University of Washington Press).

Benjamin, W. (2002) *The Arcades Project*, (eds) R. Tiedemann, H.T. Eiland and K. McLaughlin (Cambridge: Belknap Press of Harvard University Press).

Bolland, O.N. (2001) *The Politics of Labour in the British Caribbean* (Kingston and London: Ian Randle and James Currey).

Bond, J. (1947) *Field Guide to Birds of the West Indies: a Guide to all the Species of Birds Known from the Greater Antilles, Lesser Antilles and Bahama Islands* (New York: The Macmillan Company).

Cresswell, T. (2006) *On the Move: Mobility in the Modern Western World* (London: Routledge).

Davis, C.E. (1989) *Jamaica in the World Aluminum Industry, 1838–1973, Vol. I.* (Kingston and London: Jamaica Bauxite Institute).

Dodman, D.R. (2008) Making a living: small-scale metal smelting as a livelihood strategy in Kingston, Jamaica. In: J. Rivke (ed.) *The Caribbean City* (Kingston: Ian Randle, pp. 227–46).

Garn, A. (ed.) (2007) *Exit to Tomorrow: World's Fair Architecture, Design, Fashion, 1933–2005* (New York: Universe Publishing).

Gitlitz, J. (2002) *Trashed Cans: The Global Environmental Impacts of Aluminum Can Wasting in America* (Arlington: Container Recycling Institute).

Grove, R. (1995) *Green Imperialism: Colonial Expansion, Tropical Island Edens, and the Origins of Environmentalism, 1600–1860* (Cambridge: Cambridge University Press).

Hachez-Leroy, F. (1999) *L'Aluminum Français: L'invention d'un Marché, 1911–1983* (Paris: CNRS Editions, avec le concours de l'Institut pout l'histoire de l'aluminum).

Harris, W. (2009) The mystery of timelessness. In: K. Gyssels and B. Ledent (eds) *The Caribbean Writer as Warrior of the Imaginary* (Amsterdam: Rodopi).

Hays, K.M. and Miller, D. (eds) (2008) *Buckminster Fuller: Starting with the Universe* (New Haven and London, New York: Whitney Museum of American Art in Association with Yale University Press).

Horne, G. (2007) *Cold War in a Hot Zone: The United States Confronts Labor and Independence Struggles in the British West Indies* (Philadelphia: Temple University Press).

Lash, S. and Urry, J. (1994) *Economies of Signs and Space* (London: Sage).

Massey, D. (1993) Power-geometry and a progressive sense of place. In: J. Bird, B. Curtis, T. Putnam, G. Robertson and L. Tickner (eds) *Mapping the Futures: Local Cultures, Global Change* (London: Routledge).

McLeod, K. (2003) Space oddities: aliens, futurism and meaning in popular music. *Popular Music*, 22:3, 337–355.

Miller, D.P. (1996) Joseph Banks, empire, and 'centers of calculation' in late Hanoverian London. In: *Visions of Empire: Voyages, Botany, and Representations of Nature* (Cambridge: Cambridge University Press).

Mintz, S. (1986) *Sweetness and Power: The Place of Sugar in Modern History* (New York: Penguin).

Mitchell, W.J.T. (2002) *Landscape and Power*. Second edition (Chicago: University of Chicago Press).

Molotch, H. (2005) *Where Stuff Comes From: How Toasters, Toilets, Cars, Computers, and Many Other Things Come to Be As They Are* (New York and London: Routledge).

Nye, D. (1994) *American Technological Sublime* (Cambridge and London: MIT Press).

Price, R. (2007) *Travels with Tooy: History, Memory and the African American Imagination* (Chicago: University of Chicago Press).

Sheller, M. (2003) *Consuming the Caribbean: From Arawaks to Zombies* (London and New York: Routledge).

Sheller, M. (forthcoming) *Aluminum Dreams*.

Slater, C. (1996) Amazonia as edenic narrative. In: W. Cronon (ed.) *Uncommon Ground: Rethinking the Human Place in Nature* (New York: W.W. Norton).

Smith, G.D. (1988) *From Monopoly to Competition: The Transformation of Alcoa, 1888–1986* (Cambridge and New York: Cambridge University Press).

Solnit, R. (1993) *River of Shadows: Eadweard Muybridge and the Technological Wild West* (New York: Penguin).

Stephens, E.H. (1987) Minerals strategies and development: international political economy, state, class, and the role of bauxite/aluminum and copper industries in Jamaica and Peru. *Studies in Comparative International Development,* 22:3.

Stephens, E.H. and Stephens, J. (1986) *Democratic Socialism in Jamaica: The Political Movement and Social Transformation in Dependent Capitalism* (Princeton: Princeton University Press).

Thompson, K. (2006) *An Eye for the Tropics* (Durham: Duke University Press).

Trouillot, M-R. (2003) *Global Transformations: Anthropology and the Modern World* (New York: Palgrave Macmillan).

Urry, J. (2002) *The Tourist Gaze.* Second edition (London: Sage).

Wilquin, H. (2001) *Aluminum Architecture: Construction and Details* (Basel, Berlin and Boston: Birkhäuser).

Chapter 3

Visuality, "China Commodity City", and the Force of Things

Mark Jackson

... the commodity [i]s the elementary particle in the capitalist field ...

Shelton, 2007: 161

This chapter relates a cultural economy of Yiwu, China's self-styled "Commodity City" in eastern Zhejiang Province. Yiwu's peculiar, if also very dull, exhibition spaces and associated visualities are predicated on, and emergent from, the assumptions and expectations inherent to commodities, their production, and their global proliferation. These commodities and their visual spaces construct, in their place and promise, multiple social and cultural affordances both for China, and for global economic imaginaries. Materially immanent, the imaginary social affordances expressed by the cultural economy of Yiwu render the city and its spaces particularly suitable to critical examination, particularly as the city itself is built around exhibitionary imperatives of vision. The chapter argues, however, that the relationship between vision and matter in Yiwu's commodity spaces are substantive in ways beyond their simple representational registers.

Current work in human geography, political theory, and political ecology argues for a greater conceptual rigour to be given to the stuff of thought (e.g. Bennett, 2010; Connelly, 2011; Coole and Frost, 2010; Braun and Whatmore, 2010). This work seeks to "signal the constitutive nature of material processes and entities in social and political life; things of every imaginable kind constitute the common worlds that we share and the dense fabric of relations in and through which we live" (Braun and Whatmore, 2010: ix). Per se not unique, the conceptual genealogies of this critical trajectory identify the epistemic as emergent and immanent from within a material plenum of becoming, a universal vastness Bataille identified with the term "general economy" (Bataille, 1991). This chapter attempts to read the visual and the material within the context of the commodity imperative and Yiwu's urban space as a function of the constitutive onto-epistemological inter-relationships of matter and meaning.

An illustration might help to illuminate what is meant. Karen Barad (2007: 308) explains that for contemporary quantum theory, the field and particle exist in complementarity. "It is impossible to draw a sharp separation between an independent behaviour of atomic objects [particle] and their interactions with the measuring instruments, which serve to define the conditions under which the phenomena occur [field]." I would like to suggest that if we engage Barad's diagram as a means to understand the material and conceptual basis for our social lives – as Shelton gestures in the epigraph (see also 2007: 161) to illuminate capitalist commodity relations – then the visual (as one observing instrument) is itself productive of fields of socio-economic life by which commodities (i.e. material) are produced as meaningful, and vice versa. I want to argue that the visual and the material are co-constitutive in their productive relations of becoming meaningful, and to do so, the following will examine the ontological (and thus political) entanglements of matter, vision, and meaning through the entanglement of a Chinese commodity assemblage.

The focus will be on a contemporary cultural economy of consumption, the materialities and spaces of wholesale commodities, and, most importantly, the confederated conceptual and political expectations which inhere and emerge through the agencies of both the things and their spaces specific to Yiwu. Yiwu is considered by China to be one of the economic, moral, cultural, and entrepreneurial centres of contemporary "worlding".[1] Under the motto, "Face the world. Serving the country", Yiwu is promoted as a model for economic reform for a world visualized, and thus made possible, through the material magic of the commodity.

Within the context of this chapter's focus on the constitutive relationships between the visual and material, Yiwu and its urban exhibition logics are significant for several reasons. First, the city and its formative presence in China's growth predicated economic development stands as an emblem for national and social aspiration. Second, this aspiration is itself articulated through the material grammars of commodities, their globalising reach, and attractant power. Yiwu is, after all, a site which promotes itself, through a Chinese modernization imperative, as a place for global traders to come, fundamentally, to *see* goods and therein to disperse their (attributed) value around the globe. Finally, as we know, commodity forms materially structure the political practices of being and becoming critical to twenty-first-century grammars of living, both for what they make possible and for what, increasingly catastrophically, they are recognised as delimiting. Therefore, the present chapter sets as its task an examination of the performative production of a commodity imperative through the constitutive visual grammars of a highly commoditized urban space, yet one whose focus is less end-user consumption

---

1   The term "worlding" (Heidegger, 1977) simply means that a thing becomes a thing in virtue of its relationships in and through the fields of its action, and the actions by which that thing is recognized as possible. For instance, a pen is a pen and not a piece of plastic by virtue of it being used to write with, and writing is part of the conceptual field by which the pen is made possible as a pen. My use of the term "worlding" here refers to the production of a world by which commodities, particularly those "made in China", are recognized as fundamental to the development of a modern Chinese place in a global economy. The world is world in so far as commodities produced in China are fundamental to how both commodities and economic meaning are possible for an understanding of a global capitalist present.

(which is the typical focus of critical analyses), but rather productive promise *visually* constituted at the formative *beginning* of commodity chains.

Figure 3.1   Old woman and the promise of things, December 2008

Source: Image by author

Vision *is* the Thing

Much has been written about China's emergent economy and its enormous socio-spatial shifts, in particular its urbanization (e.g. Campanella, 2008; Friedmann, 2005; King, 2004; Ma and Wu, 2006; Walker and Buck, 2007; Wu, 2007; Zhang, 2001). Much has also been written on political economies of consumption, commoditization, and visuality (e.g. Bauman, 2007; Bocock, 1993; Cook et al., 2007; Goodman, Goodman and Redclift, 2010; Miles, 2010; Miller, 1987). Such work has ordinarily focused on the ideological and hegemonic disingenuity and alienation of social relations at the core of capitalist productions of life which fetishize phantasmagorias of commodification. In turn, the predominance of critical political economies of the recent development of "capitalism with Chinese characteristics" (Huang, 2008), and China's "being consumed by the practices and rhetorics of consumption" (Latham, 2006) has focused on, perhaps understandably (if also sometimes hypocritically), the scalar risks (social, environmental, developmental, and economic) of rapid socio-economic capitalization through an over-dependence on consumption based development. Critiques levelled against such changes often recommend, in Dirlik's words, the need "to contain the corrosive effects of commodification in a society where the present betrays no visible links to a past or an imagined future – except in the assertions of future power" (1996: 195). Similarly, David Harvey's well known critiques (for ex. 2007, 2010a) warn that neo-liberalized capacities, which have given rise to China's commodity production related changes, are inevitably going to perpetuate capitalist cycles of crash and crisis (2010b), never mind the additional injustices of growing inequalities and environmental catastrophe; thus, anti-capitalist imperatives are drastically needed to think the world differently than those dynamics currently shaping China's consumption related developments.

Rather than continue to rehearse such claims – despite their importance – this chapter argues the following: in order to understand how it might be possible to think the world differently, anti-capitalist imperatives need to focus beyond simply the assumption that an exercise of rational and political will can constitute the changes necessary for socio-political transformation, in China or elsewhere. In addition to the important critical recognitions expressed by critical political economies, we need also to attend, quite fundamentally (i.e. ontologically) to how collective and individual self-understanding, agency, and, indeed, collective will, are enfolded and emergent *from* the material spaces within which they operate. As Bernard Steigler notes in the introduction to his recent critique of political economy, the objects which confederate our lives "constitute an intergenerational support of memory which, as material culture, overdetermines learning and [meaningful social] activity" (2010, 9). Thus, in order to understand the spatialization of thought, politics, and social life, we need to understand how we are always already enfolded within the "materialization of experience" (*ibid*, 8). That the materialization of experience and its globally oriented promises are, in Yiwu, predicated visually renders its analysis particularly apropos a volume on the social meanings and practices of visual/material relationships.

Further, complicity with consumption is not simply ideological, for the ideological, as we know, is also embodied, experiential, unthought, and, in many key ways, affectual; it is part of the very viscerality by which we experience ourselves as selves, as comfortable, as able, as fearful, desirous, etc.; which is why it is so difficult to change habits of life. Indeed, we are summoned in and from the plenum that habituates us; we are, quite literally, *in it*, and as a part of it always already; we are relationally composed through our material worldings (Massumi, 2000). As Tim Ingold writes, we are inhabitants "through a world-in-formation rather than across its pre-formed surface" (2008, 30). Ingold was writing about wind and weather. But are the materials and visuals which populate our everydays, not, in every sense of the word, the atmospheres for our thoughts and deeds? Are they too not the worlds-in-formation through which we live? Of course they are. Given that the commodity and its spaces are primarily disciplined through the "quasi-religious" registers of the spectacle (e.g. Benjamin, 1996; Garoian and Gaudelius, 2004; Lasch, 1991; Milbank, 2008), paying attention to how commodities visually grammatize and co-constitute socio-political dynamics of expectation and participation requires addressing the visual grammars of commodities and their spaces, and crucially, as such how we inhabit and co-habit our worlds-in-formation.

Doing so requires, however, that we treat the visual not simply as a semiotic register; the visual is not removed from the material. Rather than merely being a means to access and/or deconstruct a specific *textuality* of thing and urban space (it is also that), we need to recognize the visual as such as a *material vector* which constitutes subjects, and thus by which we might interrogate how semiotic and political narratives emerge from agencies distributed across human and non-human complexes, i.e. collectives of place, form, matter, economy, necessity, and desire. The chapter proposes that if critiques beginning from the perspectives of phantasmagoria, fetishism, and ideology privilege the visual, they can sometimes do so at the expense of addressing how the social is enrolled from agential negotiations between thing and thought (Latour, 2000); the visual, therefore, is key to unpicking this material, and thus virtual, negotiation.

The account begins from the assumption that commodities, their spaces, and, fundamentally, their critical analyses, are intra-actively enfolded in co-implicated capacities to think and act (Barad, 2007; Bennett, 2010; Whatmore, 2002). Thought and matter are one and the same, literally. Critiques of commodification and the politics of their fetishism (of which the following is also a part – over-consumption, "growth is good" economic dependencies, and consequent ecological and social implications are, of course, pressing) need to attend to specifics of emergence from human–nonhuman assemblages. In the present case, the focus is an interesting, influential, and little known exemplar of Chinese urbanization and economic globalization whose forms come into sight from the minutiae and promise of things in their billions, the spaces which exhibit them, and the possibilities for life which spread from, and within, the city itself. Visuality, commodities, exhibition spaces, and their materialities, within and from Yiwu, build expectant modernities and narratives of development as disciplinary archipelagos stretched across networked, multi-mediatized, and globalized spaces. By bringing to the fore this space and its banality, and, as such, its importance, we may cohere a theoretically productive procedure for thinking a culture of the present through its common, prosaic, but worlding objects and spaces (Kracauer, 1995: 75). The intention here, unlike some contemporary accounts which show how consumers are complicit within, and define, cultures of consumption (Miles, 2010), is to suggest that reading such complicity is a matter *also* of extending social action and human will into and across the variegated structures which make possible consumption and its more often than not palpable affects.

The implication of the argument, and perhaps following on from Steigler, is the recognition that Yiwu and its exhibition city-spaces signify an historical entrance into a new material-discursive complex, what I call the "consumptive apparatus". By "consumptive apparatus" I seek to suggest, if in a preliminary way, that the commodity, as a distributed assemblage concentrated in a specific artefact, replicates *itself* (human agency is fundamentally implicate within non-human – even artefactual – agency) as the frame of social and material possibility. It is, if you like, implicate as a self-organizing emergence of things; this metaphor requires us to address a critique of the capitalist field as much through the literal and micro-spatial "stuffness" of thought and its affects, as through the macro-lenses of hegemony and ideology; the field is predicated in the particle and the particle in the field.

Visualizing Yiwu's Significance

Widely circulated on English and Chinese language television in Eastern China, a short, eleven second segment advertises Yiwu.[2] In it, an obviously "Western" visitor to the city is depicted wandering agape amidst a variety of commodity exhibitions. Male, white, young, suited, and styled as a buyer engaged in business, the visitor is escorted by Chinese guides through a number of attractive and shiny scenes; the scenes are spaces characteristic to the city. Absorbed amongst glass cabinets, he stoops to stare wordlessly, and in rapt wonder, at the numerous treasures on show: ornamental sculptures, jewellery, porcelain, silks, etc. Laughing and enchanted, he twirls as he dons a blue, dragon embroidered jacket. His minders, too, smile and exclaim encouragingly. In another scene, he and a guide stand immobile yet immersed within a graphically enhanced exhibition building; it rushes past them both in a metaphor for seemingly boundless consumptive excitation. Viewers, too, enter the visitor's visual subject position: greeted in front of an aeroplane, you and your avatar are made welcome; uniformed attendants bow or stand at attention in a gleaming business hotel; luxury car doors are opened. Everyone smiles deferentially. A voice-over intones the promise of your destination: "More fun. More opportunities. Newer city everyday. It is here. My dreamland. Yiwu, China." In the closing scene, the visitor finishes his transactions with a vigorous handshake in front of a large shopping mall-like building. Fireworks explode in a darkening sky. Yellow unspoken captions materialize the advert's conclusion: "The sea of commodities, the paradise for merchants. Yiwu, China."

Repeatedly broadcast (I saw it dozens of times during a recent visit to China), the advertisement is a much truncated version of a longer four and half minute television piece, usually broadcast only in Chinese, which extols the virtues of Yiwu as a national and global manufacturing and wholesale destination for international trade and commerce. Both adverts visually and sonically locate a modern Chinese ethos of entrepreneurial and commodity driven development

---

2   A longer version from which the short is excerpted can be seen online at, <www.youtube.com/watch?v=_iV9wtwWh5E>.

Figure 3.2   Old sock market, now disused, Yiwu, December 2008

Source: Image by author

through the symbolic register, and actual place, of Yiwu. Of course, Yiwu is not the only city so heralded in China or abroad. Shanghai, Beijing, Tianjin, Shenzen, Chongqing, Guangzhou, and the like dominate the popular imagination and much writing, scholarly and otherwise, about Chinese cities. Much less, however, is written about China's small cities, their enormous transformations, and the importance such transformations have, both for China and for shaping much of our internationally enmeshed, material and profoundly unthought everyday lives.

Not only is Yiwu a major site within China for promoting consumptive development, and, as such, an urban catalyst for aspirant narratives and their representation, it is, by some accounts (Mu, 2010), the world's largest market for small and petty commodity goods.[3] "Small commodities" means those taken-for-granteds which infuse and afford our habitats with ease, effort and expectation.

_____

3   This claim is one much trumpeted by Yiwu's sellers, agents, and on various government websites. The claim makes repeated reference to a "China Development Report" co-authored by the UN, the World Bank, and Morgan Stanley in 2005. The author has been unable to locate any such document or report to corroborate the claims. It is, of course, more than likely that Yiwu is, if not the largest small commodities market, then certainly one of the largest in the world. But, the repeated invocation in the claims of a tripartite hegemon of global capitalist authority no doubt lends a considerable credence to a claim, which as far as can be made out, is entirely viral.

They are the plenum within which the vast majority of modern life is experienced: elastic bands, mannequins, earring studs, artificial flowers, Christmas decorations, pencil sharpeners, kitchen scouring pads, the rhinestones on your Chihuahua's collar, the pen in that conference bag, the blanket you saw distributed to the Afghan refugee on television, the cheap smelly tennis ball you bought at the local pound store. The UNHRC, Carrefour, Wal-Mart, Tesco, Metro, and many other transnational organizations and corporations upon whom everyday global dependencies (conceptual, moral, and practical) are wrought have procurement centres in Yiwu; the city is probably a part of each our lives whether we know it or not. But, it is by means of its visuality, especially its exhibition driven market logic that transnational buyers (and many others) are attracted to Yiwu, and by which the city has become globally significant; indeed, it is one of *the* national and international centres for the enormous retail driven effort which shapes key experiential and cultural features of contemporary late-modernity.

Yiwu is Where?

Yiwu's urban agglomeration sits in the centre of Zhejiang Province's influential and expansive industrial zone some 300 kilometres south of Shanghai, or two hours by direct train. The production zone extends up a wide valley to Zhuji and Huangzhou in the north, south to Jinhua and Changshan, and east to Lishui. The area is dominated by tens of thousands of factories which range in size from small, one room production units to large, more stereotypical factory complexes. Yiwu's population of approximately 1.3 million people was composed, in 2008, of some 620,000 city residents, with the remainder – the majority – migrant workers. Though its influence is large, Yiwu, by Chinese standards is a very small city. Yet, the site of China's commodity trading index since 2006, Yiwu has become China's largest market and distribution centre for what the Chinese government calls "daily commodities" (Government of Zhejiang Province, 2009). Of course, given China's position in the global production economy, this also means that Yiwu is one of the world's largest, by volume, markets for small commodities, and is arguably, a central and critically important hub for the materialization of our everyday.

With a strategic initial investment of $10 million US in 1982, a series of permanent exhibition spaces, the "Zhejiang China Small Commodities City Group", were established by the government. Now numbering over twenty-four, the remit of these themed exhibitions spaces is to attract buyers and their trade. A number of unique features make this possible. First, the exhibitions themselves, and the guided access mediated by import/export companies, provide easy contact to numerous producers and sellers across an immense product range. For instance, one buyer from West Africa can purchase large, tailored volumes of inexpensive clothing, shoes, blankets, utensils, knick-knacks, cheap electronics, furniture, and farm implements (and a myriad of other things) over the course of a few days, in one location, and with relative ease. Second, the close proximity of the exhibitions and sellers to the many factories and assembly units in the region allows buyers both to customize product requirements, and to trace the commodity chain processes to the production designers and the factory floor. Traders routinely visit local factories and assembly facilities to confirm production requirements and to build relationships with manufacturers. Third, Yiwu operates, fundamentally, on an economy of scale. Although the city is enervated by the rhetorics and imaginaries of personal consumption and pleasure, it is primarily neither a place for leisure nor private consumerism. It is a place for foreign wholesale buyers to come to design and purchase, in large quantities, small commodities which will be shipped to the buyers' home market. The more units a buyer purchases, the lower the individual cost of the unit and the cheaper the costs of production and shipping. What privately oriented retail there is – and there is, to be sure, an awful lot – is situated in proximal relationships to a larger apparatus of consumptive production and commodity promise which is both national and international in focus. Yiwu's stock in trade and very reason for being a global hub is wholesale import/export transactions.

Cultural imaginaries of commodity promise and aspirant participation have transformed the immense growth of the city's market revenues from $470,000 US, soon after initial state investment in 1982, to $2.2 billion US in 1996. Since 1996, revenues from the city's markets have continued to grow rapidly. Before the global economic downturn of 2008, the annual income of Yiwu's exhibition markets, in 2006, peaked at $10.67 billion dollars US (Zhejiang Government, 2008). These are no small numbers; but the story does not end (or begin) there.

... it is Everywhere: Fakes, Exhibitions, Containers

As an indicator of the power and magnetic quality of the commodity in the transformation of social, political, and cultural life, it is interesting to note that Yiwu and its surrounding regions are also well known to be amongst the world's largest producers and marketers of counterfeit products (Chow, 2001, 2006; Engels, 2010; Lowenstein, 2001). The fact that the city was once, and continues, if to a smaller degree, to be a global centre for "fakes" illustrates a symptom of the space negotiating its entrance into a visual culture of participation through an exteriorizing "*in*-clusion" (Sardar, 2000; Abbas, 2008:_253). As Ackbar Abbas points out, the symptom of the fake disappears when cities and economies enter into a global establishment (Abbas, 2008: 254). Whether this will happen to Yiwu remains to be seen. But, given the fact that the city is a major production site for global imaginaries through cheap commodities to often very poor regions around the world, the networked globalities which make mobilities of commodities possible, and the inequalities upon which contemporary capitalism is premised, will no doubt mean that the disappearance of fakes from Yiwu is a long way off. Thus, noting the importance of the counterfeit to Yiwu and its commodity networks is to take seriously the cultural power of the brand, its visuality, and the visuality of the commodity to traverse from a locus of significant commercial import across the globe precisely through the aspirant modernities, social forces, and intimate complexities which come to be attached to the commodity, counterfeit or not. The counterfeit, as the aspirant commodity par excellence, operates precisely through its deliberate visibility, and often its obvious illegitimacy, as a material register for performatively constituting promissory imaginaries. Indeed, their illegitimacy reveals precisely the materialization of experience which inheres through a social and promissory memory whose basis lies in the commodity as such.

At the basis of this visual symptom is a doubled entanglement of commodity force (exemplified in the counterfeit) and permanent exhibition which house the mostly legitimate products, but also a number of purveyors of counterfeits.[4] The commodity thus architects the spaces of its visuality, and in so far as the wholesale exhibition markets model themselves on shopping malls in name and character ("Jinfuyuan Jewelryplaz"[sic], "Fashion Lady Shopping Mall", "Old Beimen New World Shopping Centre", etc.), the spaces architect the visual grammars by which socio-economic possibility circulates. Indeed, just as Yiwu has modelled an economic success on the hybrid marriage of exhibition and shopping mall, so now Yiwu and its emphasis on commoditized visuality has travelled as material exemplar for urban development.

Yiwu has built strategic economic and urban development linkages in developing markets worldwide. These continue the productive and visual power of the commodity to promise an aspirant modernity through exhibition. The following are some examples modelled explicitly on the Yiwu's exhibition premise which have been built, or are in the process of being built: China Trading City in Cambodia; China Trading City in Brazil; Dragon Mart and Chinese Commodity Fair in the United Arab Emirates; China Door in South Africa; China Town in Italy; China Merchants Trade City in Hungary; and, China Trade City in Poland. Similar initiatives, explicitly attempting to recreate the Yiwu model of exhibition trading for commodity wholesale were attempted with local state backing and Chinese foreign direct investment in Sweden (China Europe Business and Exhibition Centre, Kalmar) and in The Netherlands (Chinamex Europe Trade and Exhibition Center, Schiphol) only to fail. But, it is telling that those Yiwu modelled initiatives which have succeeded, even perhaps that of the Italian example, have done so in emerging capitalist markets where relatively inexpensive commodities, and the promises they inhere for participation in practices and lifestyles of late capitalist modernity, in places like Brazil, Cambodia, Poland, and South Africa. The aspirant materializations of experience communicated through the visual grammars which circulate from and through Yiwu are emergent from a commodity predicated

---

4   In the dim corners of the city's largest and most important permanent exhibition, Futian Market, I came across, for instance, "dadsolid", a wholesaler selling "Dunhill" copied luggage, and "Werwilson" which sold "Wilson" copied sports gear. There were many more such examples in the streets of Yiwu itself and in its exhibitions.

material culture which understands itself as having global reach and economic meaning. Where they failed (Sweden and The Netherlands), they failed precisely because the urban and socio-material infrastructures which articulate around and through commodity promise and production in Yiwu were sedimented in quite different ways, and thus were much less fluid (Sandberg, pers. comm.; see also, Jansson and Sandberg, 2008). As with the fake, so with the exhibition.

Part of the fluidity (beyond financial) which supports the possibility of a consumption based development to articulate around the "Made in China" production is, of course, the shipping container, which make possible the efficient and inexpensive flow of goods around the world. (It is the shipping container which allows the global travel of the Yiwu based model of exhibition capital development.) Indeed, shipping containers have come to symbolize Yiwu, so much so that at a popular visual art exhibition in Beijing in 2006, the Chinese artist, Liu Jianhua, displayed a mixed media installation called "Yiwu Survey (2006)" complete with a scale model of a red shipping container disgorging its jumbled contents – plastic toys, rice cookers, dustpans, calculators, etc. – in piles on the gallery floor. An interview with a shipping agent in Yiwu revealed that a common parlance amongst agents is that of customer/container ratios; anything less than one customer to one container is slow business. Thus, the container and its standardized, boxed contents signify a competitive capacity to participate in the horizon of networked modernity, a participation which subtends the virtual promise contained within the boxes themselves.

What is key in the examples of counterfeit, shipping container and global exhibition development is a constitutive complementary relationship between the matters of visuality and the architecting of space. The force of the thing which articulates desire as need through, for instance, the counterfeit is made possible on a large scale by the efficient mobilization of material flows and distribution. As such, the very model of urban space predicated on what neo-liberalised and globalized capital made possible for Yiwu has come to travel (in limited ways) as thinkable for disparate and commodity hungry regions around the globe. The field of socio-economic relations is traceable through, and only because of, the force of the specific thing to exercise itself as a field. If a commodity is a concentrated locus or node which holds together and makes possible the analysis of a vast assemblage of social, material, political and personal relations, then Yiwu, likewise, is a concentrated node by which to analyze the topologies of urban space which articulate consumption as consumption of the spectacle. Yiwu, therefore, is part of producing an iconography of imagination – a framing and thus worlding – by which the fantasy space of a promissory urban and global modernity is visualized through the material promise of the commodity as economic potential. This materialized picturing of capitalist life-world is such that what we see in contemporary Yiwu is now, as Abbas argues, not the commodification of culture, but the "acculturation of the commodity" (2008: 259). The polarities of analytic force have tipped such that what now needs critically attending is less an analysis of ideology than of the political ontologies immanent in the thing itself. The thing and its spaces (for increasingly they are one and the same) are agentially potent assemblages through which culture is produced as force field. Let us now turn then to the exhibition spaces and the things themselves.

Exhibition as World: World as Exhibition

Yiwu has become defined by its now more than twenty-four themed exhibition markets which constellate the city. These specialized exhibition markets are each devoted to displaying and selling all manner of things. There are markets for: moulds, building supplies, bedding, socks, belts, wood, furniture, jewellery, clothes, etc. The city's commodity imperative is dominated, however, by a four million square metre market referred to as "International Trade City", or more commonly, "Futian Market". The Futian complex hosts tens of thousands of shops under one roof and sells some 1.7 million different commodities to buyers from around the world, who at its peak in 2007 are said to have numbered at approximately 40,000 visitors per day (Zhejiang Government, 2009). As a key locus of object-desire the Futian Market architects an imaginary and emergent global futurity through a faith in a material fantasy that demands visual exhibition.

It is important to note that the exhibition space and the socio-political assemblages to which it is tied, and which it makes possible, are not new to China (nor unique to Yiwu; China has a considerable fascination with the economic and cultural power of the exhibition – consider the cultural and political importance to China of the Expo 2010 Shanghai). Permanent and temporary exhibitions feature across the considerable history of China's commodity and material landscapes, and articulate, as they did for Imperial Britain, the United States and Europe, a participation in early modern state building and globalization (Dikötter, 2007; Yeh, 2007). While exhibitions have long colonial and nationalist histories (e.g. Bennett, 1988; Hoffenberg, 2001; Mitchell, 1989; Rydell, 1984), they began in China with the Nanyang Exposition of February 1910, which was the first nationally significant exhibition (Fernsebner, 2006). Nanyang served as an event imagined as both classroom for national development (as most exhibitions were), and as a new kind of spectacle that could accomplish the pedagogical aim of presenting the nation as a competitive force resistant to, and distinct from the colonialist traps of a global industrial economy (*ibid* 100). Today, Yiwu returns through this colonial trajectory but with differences significant both for the disciplinary work of the visual technology of the exhibition, and importantly for how the spectacle, now rather banal, works contemporaneously.

The International Market is a four story permanent market open every day of the year to exhibit and sell every imaginable form of consumer and household good. The building is organized according to commodity theme with the first several hundred booths beginning with number one devoted to artificial flowers. When I left Yiwu in December of 2008, booth number No. 62-083 in the newly opened north quarter of Futian, sold pantyhose. The noise of construction next to this last booth signalled the continuing expansion of the market north and east. The corridors of the market, all identical and without ornamentation, like the booths, are numbered, ordered and signposted like conventional city streets. Booths themselves are navigable not only by name but primarily by number and hall type: F17281, F17276, etc. Begin with shop one and move through the complex systematically – a feat that, it is boasted, would take you a year and a half if you were to stop at each stall for three minutes – and you will traverse the plenum of our material ordinary in repetition after repetition of display chambers. Dimly lit, the shops exist in a sort of suspended gloom, amidst the endless row upon row, floor upon floor, kilometre after kilometre of similarity.

It is a spleen-ic[5] experience being in Futian Market. Familiar markers to those of more personal consumption palaces are present – glass, mirrors, marble floors, echoing sounds, isolated wanderers of sparse halls, boredom, stuff, endless stuff. But it is certainly also uncanny. There is no music, light is kept to a minimum, dust and monotony pervade the entirety of the space. Futian, despite being described as "a paradise and ocean for shoppers", and despite being a centre for commodity consumption, is not a fantasy constructing space. A strange amalgam of various market and exhibition layers, it is a space suspended in faith to the summons of the material fantasy. The material beckons with its own agency; the apparatus thus forces.

The commodity city participates quite self-consciously in Orientalizing itself as an Asian market. At the south entrance to what would have been the original entrance when the space first opened, a large marble mural narrates the market through the tropes of visiting traders from around the world; it is almost as though Yiwu wants to construct itself as at the end of the Silk Road. Stylized, two dimensional Middle Eastern figures, camels, and goods caravans are seen interacting with traditionally clad Chinese figures. Young women serenade the scene with folk instruments. In the distance, signified in the uppermost regions of the mural, junks full of sail ply the waters of early modern globalizing commerce. The scene hybridizes temporalities and spatialities, re-invents histories of trade, and refigures, for the present, ancient globalizations. Bored teenagers, migrant workers employed in the stalls, dressed in skull and crossbones scarves smoke under the mural during breaks. Spatially not unlike a typical mall then, Futian invites similar embodiments and expectations: a space of boredom before the call of the material, and therein the fetish; indeed, it is the commodity–human nexus which is agentially operative (Bennett, 2010: 38).

Part mall, part exhibition, part market, Futian reflects, as repeatedly mentioned now, a larger spatial dynamic operative throughout the city itself. Yiwu has many

---

5 "Spleenic" invokes here the meaning Baudelaire assigned it in his atmospheric collection on Parisian modernity, *Les Fleurs du mal*, an ennui of things, and quiet desperation or disgust of life (Waldrup, 2006: xx). Walter Benjamin invoked Baudelaire's use of the term to characterize his impressions of the Parisian arcades, and its atmospherics became central to how the arcades embodied both the fossilized memory and ruin of modernity.

such enclosed markets devoted to exhibition and display, although all on smaller scales. Binwang Market specializes in cheap ready-made clothes, luggage and bedding. There are markets for socks and hosiery, non-staple foods, bathroom and toilets, wallpaper and paint, door handle and fixtures, lighting, furniture, small engines and appliances, a car and motorcycle market, a moulds market, jewellery and fashion ornament market, agricultural products, etc. Interestingly, beyond these territorially defined markets, the city has devoted specialized streets and areas to commodity production, exhibition and consumption; I walked through whole sections of the city devoted variously to auto-goods and accessories, scarves and hats, lingerie, photo frames, glass and resin ornaments, belts, shoes, threads and ribbons, calendars, lighters, mannequins, and Christmas ornaments (just over 70 per cent of the world's Christmas ornaments are made and sold in Yiwu). Markets and exhibitions become streets, blending arcades with the signs of the city and the domestic, weaving the physical with highly digitized environments, to produce heterotopic spaces which themselves become animate in multiple forms of display, to be taken up as markers and productions of consumptive identity, bought as phantasms of the immaterial, and which will, as active agents in de-subjectifications, disappear into the background of the banal and everyday. In places like Yiwu our material world lies in pregnant wait a filled with the tensions of future promise and catastrophe realized, or not, a tension humming in the material itself.

Yiwu and its exhibition spaces subtend the banalities of our everyday; it is a space underneath, before, but integral to, in animated suspension towards becoming everyday. It and its commodities are a part of that folded or doubled social space wherein the imaginary force of the symbolic is most raw, where they are yet to enter an individualizing subject space, but precisely because of that, intensely powerful, for they inhere the imagined possibilities of a diffused production; they are signifiers of the virtual in the process of becoming and promising the actual (Deleuze, 2007: 148–152). The hegemonies of their form and their exhibition space condense emergent possibilities that articulate the world through a material dreamscape of infinitely repeated variability, but a variability contained and transmitted through the commodity condensate. The exhibition is the world, for what "world" means from the perspective of Yiwu and its inter-related scales is defined by the visual promise of its contents and possibilities – the exhibition as world.

The Consumptive Apparatus

On the night of the 15th of December 2008, I was in a hotel room in Yiwu watching TV. CCTV, China Central Television's Channels 1 and 3 were broadcasting a special concert in celebration of the thirty-year anniversary of "opening up" (*kaifeng*) and "market reforms" (*shichang gaige*). The concert was a visual and aural spectacle, one in the long tradition of socialist events celebrating the nation and its people.

Broadcast from the recently opened National Centre of the Performing Arts across the street from the People's Great Hall in Tianamen Square, the concert featured choirs and bands of the Chinese military. The one-hundred and sixty person strong choir and huge orchestras were fronted by a trio of singers from each of the military wings, each stalwarts of a socialist imaginary: beautiful, clear eyed, fit, articulate, seemingly unquestioning, straight backed and vigorously smiling. They sang songs in praise of the nation, its military, and, on this, the thirtieth anniversary of the breathless economic reforms that have transformed the country, the new China. The concert was one event in a series of events in cities around the country, and in various media, in celebration of economic reform and opening up. On the 15th, the audience for the concert was largely made up of counterparts from the military, and, of course, a national audience of millions.

While the concert was visually spectacular for its precision, efficiency, technical virtuosity and unflinching jingoism, it caught my eye in another way. Behind the choir, on a huge screen, images of the new, urban China – rural China was glaring in its absence – its newly idealized consumer citizens, and its architects of the economic and social change swept by as the music soared: skyscrapers, planes, industry, trade, tanks, factories, and new futuristically generated CGI cityscapes. These images were complete with idealized green suburbs, gleaming, technically advanced architectures yet to be built, and, notably to my naive eyes, marketing images of cavorting shoppers and consumers, complete with branded shopping bags, impossible, carefree smiles, and the usual affectual triggers of the consumptive apparatus. Revolutionary rural workers had become revolutionary consumers. The socialist aesthetics was acculturated to the commodity fetish, but the social and collective production was perhaps less one of creating an ideal citizen subject, but of rather that of meeting the now global imperative, and so being modern.

But what also struck me was that many of the images displayed during the concert were taken directly from the same promotional video for the city of Yiwu with which I began. I had seen this promo for the city of Yiwu run interminably on both the English and Chinese channels of CCTV since arriving in China. The Chinese language version was over seven minutes long, but the most common English version was edited down to about fifteen seconds – a short punch of repeated affect rather than a sustained argument. What became clear is that Yiwu is an urban space central to the state's signifying efforts to represent the imaginary capacity of emergent Chinese modernity. CCTV 9 began its three week coverage of the thirtieth anniversary of reform and opening up with a story about the "economic miracle" of Yiwu.

Yiwu and its diverse visualities, through commodities, exhibitions spaces, counterfeits, shipping containers, globalized markets, television adverts, state celebrations – assembles/presents/depicts itself as part of a "consumptive apparatus". This apparatus is a heterogeneous network (a *dispositif*) whose strategic function emerges at the intersection of emergent power relations and relations of economic knowledge (Agamben, 2009), which for our purposes are articulated through the contemporary accumulation and proliferation of the visual and material grammars of the commodity. It is an any-space-whatever, to adapt Deleuze's phrase, whose "virtual [i.e. its imaginaries, its dreamworlds] grasps the pure locus of the possible" (1986: 109) through the world framed as commodity. Commodities thus are agentially pregnant *and* active, in producing the conditions for the possibility of subjective engagement. They are productive nutrients of social, cultural, and economic potentiality which make up the flowing global networks and fields of the everyday, and which, as such, co-constitute them as material, moral, and possible. They are the means through which flow the continuities and discontinuities of experience which resolve in the everyday because of the always already relations of visuality and matter through which we inhabit the world as possibility.

Conclusion

We could trace a history of visualization apparatuses and their architectures. Olalquiaga (2005/6) and Stafford (1999) distinguish between the *Wunderkammern* and cabinets of curiosity. *Wunderkammern* were those Renaissance rooms where objects of puzzlement, awe, and wonder were encyclopaedically displayed both as fantastic scenarios of the divine and as theatres to classical empiricism. Cabinets of curiosity emerged soon after with the more analytic desire to grasp and control that mystery, to turn wonder to knowledge through methods of scientific investigation. The museum and, importantly, the exhibition quickly followed as nineteenth-century vehicles for inscribing and disciplining cultural technologies of power as material signifiers of progress, with capital and science as the great co-ordinators. The exhibitionary complex, as Tony Bennett (1988) terms it, had, as its organizational rubric, a disciplinary production, and thus administration, of a public.

We have entered another, possibly, fourth visual complex, a consumptive apparatus, whereby the rubric of public is but an epi-phenomenon of the topological fluidities of commodity logics and their economic architectures. This apparatus is fundamentally indifferent to the work of subjectification.[6] Rather, it is part of a concern to replicate its own economically defined world whose visual, and thus embodied, affect is that of the phantasmagoric. Subjects, and thus possible publics, are no longer pre-supposed by the consumptive apparatus; the apparatus replicates itself in an almost incomprehensively vast assemblage as the frame of governmental and material possibility. Yiwu is interesting for it is but one historically recent site (if not unique, then an interesting one nonetheless) built specifically in light of this fourth visual/material apparatus, wherein the material thingness is part of the world replicating itself. It is not the case that previous apparatuses were no less entangled. Of course, they were. "Discursive practices

---

6   Benjamin might term this indifference the dream sleep or slumber of the modern.

are the material conditions that define what counts as meaningful" (Barad, 2007: 63, emphasis added). But in discerning a future for a critical politics of the present we must proceed through, literally, the architectures of the visible to understand our own enmeshment in the force of things. Only by attending to the micro-scopic affordances of these things (walking through them, picking them up, feeling them, seeing them) can we begin to shape a different force of things, and so meaning.

References

Abbas, A. (2008) Faking globalization. *Other Cities, Other Worlds: Urban Imaginaries in a Globalizing Age*. In: A. Huyssen (ed.) (Durham: Duke University Press, pp. 243–264).

Agamben, G. (2009) *What is an Apparatus? And Other Essays*. Trans. D. Kishik and S. Pedatella (Stanford: Stanford University Press).

Barad, K. (2007) *Meeting the Universe Halfway: Quantum Physics and the Entanglement of Matter and Meaning* (Durham: Duke University Press).

Bataille, G. (1991) *The Accursed Share Vol. 1: An Essay on General Economy* (New York: Zone).

Bauman, Z. (2007) *Consuming Life* (Cambridge: Polity).

Benjamin, W. (1996) Capitalism as religion. In: M. Bullock and M.W. Jennings (eds) *Selected Writings, Vol. 1, 1913–1926* (Cambridge: Harvard University Press).

Bennett, J. (2010) *Vibrant Matter: A Political Ecology of Things* (Durham and London: Duke University Press).

Bennett, T. (1988) The exhibitionary complex. *New Formations*, 4: Spring, 73–102.

Bocock, R. (1993) *Consumption* (London: Routledge).

Braun, B. and Whatmore, S. (eds) (2010) *Political Matter: Technoscience, Democracy and Public Life* (Minneapolis: University of Minnesota Press).

Campanella, T.J. (2008) *The Concrete Dragon: China's Urban Revolution and What it Means for the World* (New York: Princeton Architectural Press).

Chow, D.C.K. (2000) Counterfeiting in the People's Republic of China. *Washington University Law Quarterly*, 78:1, 1–57.

Chow, D.C.K. (2006) Why China does not take commercial piracy seriously. *Ohio Northern University Law Review*, 32:2, 203–225.

Connolly, W. (2011) *A World of Becoming* (Durham: Duke University Press).

Cook, I.J., Evans, J., Griffiths, H., Morris, R. and Wrathmell, S. (2007) It's more than just what it is: defetishising commodities, expanding fields, mobilising change … . *Geoforum*, 38:1, 113–1126.

Coole, D. and Frost, S. (eds) (2010) *New Materialisms: Ontology, Agency, and Politics* (Durham: Duke University Press).

Deleuze, G. (1986) *Cinema 1: The Movement-Image* (Minneapolis: University of Minnesota Press).

Deleuze, G. (2007) The actual and the virtual. In: *Dialogues II*. Trans. C. Parnet (ed.) (New York: Columbia University Press, pp. 148–152).

Dikötter, F. (2007) *Things Modern: Material Culture and Everyday Life in China* (London: C. Hurst & Co.).

Dirlik, A. (1996) Looking backwards in an age of global capital: thoughts on history in third world cultural criticism. In: X. Tang and S. Snyder (eds) *Pursuit of Contemporary East Asian Culture* (Boulder: Westview Press, pp. 183–215).

Engels, S. (2010) Counterfeiting and piracy: the industry perspective. *Journal of Intellectual Property Law & Practice*, 5:5, 327–331.

Fernsebner, S.R. (2006) Objects, spectacle, and a nation on display at the Nanyang exposition of 1910. *Late Imperial China*, 27:2, 99–124.

Friedmann, J. (2005) *China's Urban Transition* (Minneapolis: University of Minnesota Press).

Garoian, C.R. and Gaudelius, Y.M. (2004) The spectacle of visual culture. *Studies in Art Education*, 45:4, 298–312.

Goodman, M., Goodman, D. and Redclift, M. (eds) (2010) *Consuming Space: Placing Consumption in Perspective* (Farnham: Ashgate).

Government of China (2008) *Zheijiang Statistical Yearbook* (China Statistics Press).

Harvey, D. (2007) *Spaces of Global Capitalism* (London and New York: Verso).

Harvey, D. (2010a) *The Enigma of Capital and the Crises of Capitalism* (London: Profile Books).

Harvey, D. (2010b) Organizing for the anti-capitalist transition. <www.davidharvey. org> Accessed 10.11.10.

Heidegger, M. (1977) The question concerning technology. *The Question Concerning Technology and Other Essays*. Trans. W. Lovitt (New York: Harper, pp. 3–36).

Heidegger, M. (1977) The age of the world picture. In: *The Question Concerning Technology and Other Essays*. Trans. W. Lovitt. (New York: Harper, pp. 115–154).

Hoffenberg, P.H. (2001) *An Empire on Display: English, Indian, and Australian Exhibitions from the Chrystal Palace to the Great War* (Berkeley: University of California Press).

Ingold, T. (2008) Earth, sky, wind, and weather. *Wind, Life, Health* (Oxford: Blackwell, pp. 17–35).

Jansson, H. and Sandberg, S. (2008) Internationalization of small and medium sized enterprises in the Baltic Sea region. *Journal of International Management*, 14:1, 65–77.

King, A. (2004) Villafication: the transformation of Chinese cities. In: *Spaces of Global Cultures: Architecture Urbanism Identity* (London and New York: Routledge, pp. 111–126).

Kracauer, S. (1995) *The Mass Ornament – Weimar Essays*. Trans. T.Y. Levin (Cambridge: Harvard University Press).

Lasch, C. (1991) *The Culture of Narcissism: American Life in an Age of Diminishing Expectations* (New York: W.W. Norton).

Latham, K. (2006) Consumption and cultural change in contemporary China. In: K. Latham, S. Thompson and J. Klein (eds) *Consuming China: Approaches to Cultural Change in Contemporary China* (London and New York: Routledge, pp. 1–21).

Latour, B. (2000) When things strike back. *British Journal of Sociology*, 51:1, 107–23.

Lowenstein, A.B. (2001) Chinese fake-out. *Foreign Policy*, March/April 2001.

Ma, L.J.C. and Wu, F. (2006) Transforming China's globalizing cities. *Habitat International* urbanization in China. A special issue in two parts, 30: 2, 191–198.

Massumi, B. (2000) Too blue: colour patch for an expanded empiricism. *Cultural Studies*, 14:2, 177–226.

Milbank, J. (2008) Paul against biopolitics. *Theory Culture Society*, 25:7–8, 125–172.

Miles, S. (2010) *Spaces for Consumption* (London: Sage).

Miller, D. (1987) *Material Culture and Mass Consumption* (Oxford: Basil Blackwell).

Mitchell, T. (1988) *Colonizing Egypt* (Berkeley: University of California Press).

Mu, G. (2010) The Yiwu model of China's exhibition economy. *Provincial China*, 2:1, 91–115.

Ollalquiaga, C. (2005/06) Object lesson/transitional object. *Cabinet*, 20, Winter, <www.cabinetmagazine.org/issues/20/olalquiaga.php>. Accessed 10.02.2011.

Rabinow, P. (2008) *Marking Time: On the Anthropology of the Contemporary* (Princeton and Oxford: Princeton University Press).

Robinson, J. (2006) *Ordinary Cities: Beyond Modernity and Development* (London and New York: Routledge).

Rydell, R. (1985) *All the World's a Fair: Visions of Empire at American International Expositions, 1876–1916* (Chicago: University of Chicago Press).

Sardar, Z. (2000) *The Consumption of Kuala Lumpur* (London: Reaktion).

Shelton, A. (2007) *Dreamworlds of Alabama* (Minneapolis and London: University of Minnesota).

Stafford, B. (1999) *Artful Science Enlightenment Entertainment and the Eclipse of Visual Education* (Cambridge: MIT Press).

Steigler, B. (2010) *For a New Critique of Political Economy*. Trans. D. Ross (Cambridge: Polity).

Waldrup, K. (2006) *Translator's Introduction to Les Fleurs du Mal' in The Flowers of Evil by Charles Baudelaire* (Middleton: Wesleyan University Press).

Walker, R. and Buck, D. (2007) The Chinese road: cities in the transition to capitalism. *New Left Review*, 46: 39–66.

Whatmore, S. (2002) *Hybrid Geographies: Natures Cultures Spaces* (London and New York: Routledge).

Wu, F. (ed.) (2007) *China's Emerging Cities: The Making of New Urbanism* (London and New York: Routledge).

Yeh, W.-H. (2007) *Shanghai Splendour: Economic Sentiments and the Making of Modern China, 1843–1949* (Berkeley and London: University of California Press).

Zhang, L. (2001) *Strangers in the City: Reconfigurations of Space, Power, and Social Networks within China's Floating Population* (Stanford: Stanford University Press).

Zhejiang Government <www.zj.gov.cn/zjforeign/english/node493/node499/userobject1ai5840.html>. Accessed 09.08.2009.

Video

Yiwu Video. www.youtube.com/watch?v=_iV9wtwWh5E. Accessed: 28.03.2011.

Chapter 4

Tristes Entropique: Steel, Ships and Time Images for Late Modernity

Mike Crang

Never forget: Decay is inherent in all composite things or compounded phenomena.

Gautama Buddha's last words

Life, in [Bergson's] philosophy, is a continuous stream, in which all divisions are artificial and unreal. Separate things, beginnings and endings, are mere convenient fictions: there is only smooth, unbroken transition ... All our thinking consists of convenient fictions, imaginary congealings of the stream: reality flows on in spite of all our fictions.

Bertrand Russell, 1918, 22

The masthead quotations exemplify a long history of thinking about materiality and temporality through flux and flow. The question, then, is how do we envision such incessant movement? How do our images capture flow? Bergson rejected the 'cinematic illusion' as a solution since the moving image was actually composed of stills. How can an image convey movement in itself, and not just across a series? Michel Serres derives this sort of materiality from the physics of Lucretius that sees the apparent stability of the world as an illusion caused by the turbidity of incessant flows. In this vision atoms initially fall like rain through a void, in a parallel laminar flow that then is disturbed, with a smallest of spins or swerves (the Clinamen of Epicurus) that produce forms (Bennett, 2001: 100) as the flow becomes turbulent, 'leading to vortices in which the atoms combine to form a quasi-stable order' forming a world out of the myriad combinations of atoms arising from chance encounters and remaining held by the regular movement of the flow, setting 'aside the principles and habits of thinking in terms of solids and treats atoms as the condition for a theory of flow' (Webb, 2006: 127). The world is constantly in flow, just some of it at a very slow rate, and full then of nonorganic life, as De Landa (1992) argued. Such an approach highlights not things moving through empty space, but the world as becoming-things.

Jane Bennett (2010) takes just this approach to ask about the life of metal. Whereas steel has conventionally been scripted as the epitome of fixed, static and obdurate material, she wants to reinscribe a vibrancy within it to challenge what, after Gilbert Simonden, she calls the 'hylomorphic' model of mechanistic world reliant on external actors providing formative and innovative possibilities. Instead she uses Lucretian atomism to figure the latent topological tendencies that do not merely resist or respond to external forces but endeavour to express themselves. Following Serres there is a 'conception of space in atomism is at once topological and radically material; an unusual combination that frees it to describe a series of non-metric relations between spatial, temporal, and discursive localities' (Webb, 2006: 126). The material things are not prior to discourse, not prior to agency, but are part of both. This is a vision of metal as 'both material and creative, rather than mechanical and equilibrium-maintaining,' that draws on Deleuze and Guattari to see the potential of becoming.

The focus of this chapter is though a negative becoming, or a sense of productivity that includes failure, disassembly and destruction. Following the acknowledgement of the crystalline internal irregularities of steel sees them leading to failures as well as strengths; imperfections in crystalline structures produce both sharpness and brittleness (de Landa, 2006). Rust, breakdown and destruction are immanent propensities of, not exceptions to, the normal state. Indeed one might say that Bennett's history, though it celebrates the creative knowledge of metal workers and craftsmen who make things, is shorn of the knowledge and labour of those who struggle to maintain things – those who fight the long defeat against wear and tear – although Bennett correctly suggests that apparent solidity is the result of the slowness of material process and the brevity of our observation. Here then we come to the link of the flow and event of things and the moment of visualization – how does that event in its brief happening and lingering image link to the slow happening of inorganic life? It may be worth connecting that slow happening to the notions of entropy – that matter heads towards increasing levels of disorganization – that is, it is things are generally unbecoming. Even the great steel sculptures of artists like Richard Serra speak to this unbecoming since whilst 'often monumental in scale, ... [seem] for all their monumentality, to be just as much about rust' (Maskit, 2007: 329). This directionality given to matter is perhaps most influentially worked through in the work of Robert Smithson.

Though they do not make the connection to wider vocabularies of vitalist materialisms, Kathryn Yusoff and Jennifer Gabrys (2006: 447) have recently highlighted how for Smithson 'the mobilization of matter involves its inevitable decay', as well as how he emphasised that 'solids are particles built up around flux.' Robert Smithson argued that separate 'things,' 'forms,' 'objects,' 'shapes,' were 'mere convenient fictions: there is only an uncertain disintegrating order that transcends the limits of rational separations' (1996, 112). Much of his work was about ways of visualizing temporality, transience and entropy, highlighted by the 'entropic architecture or a de-architecturization' in his 'partly buried shed' in Kent State University's grounds, where he put 20-loads of soil on top of a wooden shed till it partially caved in, and now nearly forty years later what remains is a wooded hillock. He spoke evocatively of Rome as 'like a big scrap heap of antiquities.' Noting that 'America doesn't have that kind of historical background of debris,' he then went to discover such debris in photographic projects like *Monuments of Passaic*. There the industrial effluent pipes of the Jersey shore become a fountain monument, and the industrialized coast gets mapped as a kind of archaeological park of ruins 'yet to be,' where industry is already set up as though in a picture of future failure – rendered visible and exposed (Smithson, 1996: 70–2).

If we acknowledge that 'many operations of knowledge are already at work within the objects of the world' (Serres and Hallward, 2003: 231), it may be we ask whether in representing the world there are not some forms and modes that partake more readily of this perspective on materiality in motion. One might look at the attentive material semiosis in the floral sculptures of Anya Gallaccio, whose transformation by nature emphasises unruly emergence and decay, where nature overwhelms human material as abject and abandoned materials breach symbolic orders, or whose ice sculptures show the inevitable transience of form (Rugg, 2000; 2007). Another art work in tandem with nature might be Jenifer Wightman's *Winogradsky Rothko*. This combines the work of soil scientist Winogradsky in using a colour sensitive nutrient column to observe bacteria with the large abstract colour formats of a Rothko painting. The result is a large 'picture' behind glass that is actually a nutrient field colonized by waves of bacteria, each proliferating until they exhaust a section and then are in turn colonized by the next wave, spreading waves of colour over the piece. In this way,

> living organisms manufacturing the pigment are simultaneously the subject and substance of 'painterly' objectification – both object and medium, both a work of art itself and a working of autopoiesis ... What it is in its becoming is what it means (Wightman, 2008: 312).

In relation to the unwelcome vibrancy of metals, consider Mel Chin's (1989) *Revival Field* of alpine pennycress, planted to draw toxins out of a waste dump in passive remediation (Miles, 2002: 86). *Revival Field* can be seen as an attempt to displace both the agency of the sculptor and the usual notion of humans transforming nature, instead to suggest that humans and plants might recreate nature mutually from the detritus of industry.

Photography seems mechanically to stand apart as a reproduction of, not participant in, these vitalities. Yet photography plays upon time and allows the play of time to become apparent. Walter Benjamin (1978) pointed to the way photography can allow us to visualize the very small or apprehend the very fast through motion capture. What then if we turn that gaze on the flows of materials and ask how might we visualize the very large and very slow? If we look too at the preservative function of photography, whereby 'violently stopping the flow of time, [photographs] introduced a *memento mori* into visual experience' (Jay, 1993: 135), or at what Sontag (1977: 69) famously called its special aptitude for the injuries of time, then we might expect photography to be adept at highlighting temporalities of emergence but especially those of unbecoming. This association of photography with decay, dissipation and death is very much the common move in high art criticism – often heavily influenced by Barthe's melancholy work of mourning prompted by a misremembered picture of a lost time (Olin, 2002). However, that linkage is not inevitable. The frozen moments of time in John Pfahl's

The Very Rich Hours of a Compost Pile speak to decay, for sure, but also rebirth (Pauli, 2003a: 24) invoking a sense of cyclical time and natural vitality. The work of Jem Southam (2000) offers an entropic sense of dissipation and disorganization in his studies of the unpredictable falls of rocks and their erosion. But his repeated revisiting of sites also plays on cyclical time, from dew ponds that fill and empty with the weather, changing from pond to hollow and back, to the flow of tides. His series offers no predetermined direction for a world always becoming, where the flow of time also has eddies, surges and pools. As Millar (2000) argues, the series of pictures adds up to more than instants, but invokes the flow of time in an open series of smoothed timespaces. The focus though is upon the instability behind apparent permanence:

> From the dramatic as well and minor disturbances he describes at the sites, each with its own emphasis and structure, Southam unveils a land that is unstable and unpredictable, constantly moving through different states and at varying speeds. These entropic processes are the underlying 'subject' of his pictures ... For Southam the earth's instability and degenerative processes suggest other forms of upheaval and uncertainty, both social and personal, and again both dramatic and more insistent (Chandler, 2000).

We are then concerned with thinking about the creation of 'time images' that see objects as temporary stabilizations of things and relations, as coming in to being and as coming apart. The flows and connections of their material semiosis – linking discursive, imaginative and material registers – leads me to think about process of disassembly, destruction and wasting.

To provide a focus then I will take two very different renderings of the same wasting – and the life of steel. The wasting in question is the breaking or, as the industry likes to call it, recycling, of old ships on Bangladesh's beaches. On these beaches, low paid workers with few tools break up around a third of all the ships scrapped on the planet – and another half are broken up in the same way elsewhere in South Asia. A site of truly epic labour and destruction:

> At low tide, this vast mud flat at the northern corner of the Bay of Bengal – part of the Ganges/Padma Delta, the world's largest river delta – presents an almost apocalyptic scene. Here, on a 25-kilometre stretch of beach near the town of Sitakunda [north of Chittagong], oil tankers, passenger liners and fishing boats dot the dark, oil-slicked terrain like beached leviathans, the monolithic steel forms turning coastal idyll into industrial wasteland as they lie ready to be broken up (Bell, 2008).

What might we make of such disassembly, where one form of matter ends and yet the steel recycles into new afterlives? For this is not just the end of the line for ships, but the site that supplies up to 80 per cent of all the steel used in manufacturing new products in Bangladesh. In the rest of this chapter I want to ask how this landscape of forms disassembling into steel is figured in photographic practice. How does photography reveal, indeed revel, in the transcience it finds in this obdurate material through the grammar of the still image? This I do for a site where the discarded means of global commodity flows are broken apart, where the wastes of global consumerism are processed far from the usual attention of the world.

> These places look inhuman for their scale and for their poisons and hazards, but they are the landscapes on which most humans now depend. It may be industrial civilization is predicated on blindness and alienation, on not knowing ... [what makes] your pleasant first world urban/suburban existence impossible, for that knowledge might at least make that existence a little less pleasant (Solnit, 2007: 135).

Photography of many kinds has led the way in making Chittagong's foreshore in particular a site where the detritus of global consumerism becomes visible. Thus, imagery from the beach was chosen for the 2010 National Geographic Photography Contest Wallpaper and the Association of American Geographers 2011 picture prize winner (Figure 4.1). This scenery is paradoxically striking and hidden.

Figure 4.1   Absent witness, The Waste of the World

It is one of the strangest, most striking and frightening industrial sites in the world. It is large enough to be seen from space, but remains an open secret which few American people have even heard of, let alone seen (National Labor Committee, 2009, 6; see Figure 4.2).

That they are visible from space is evidenced on something like Google Earth where you can see the ships, in various states of dismemberment, lying scattered across the beach. As you track up and down the beach there are ships lying at different points, some suddenly foreshortened with bows cut off, some surrounded by chunks hewn from them, still others have ceased to be recognizably ships becoming just undefined chunks of material, while others have ceased to be a singular object and are scattered into fragments. Finally, as you stare at the sand there are the imprints of other keels, now long since turned into steel and rerolled into new forms. The ghostly markings of ships past. And given the datedness of the images, for all their technological and representational appeal as visual facts, they depict that which is no longer there: ships that have long since become steel and been recycled into new forms. And yet the photographic gaze seems still captivated by the death and destruction of the ships. *Der Spiegel*'s photo-essay on Chittagong's beaches speaks of 'Cemeteries of Steel' rather than cradles of rebirth where the lives of new products start. If the foreshore is a place where global material flows become visible it is also a place that has fascinated photographers, with the Panos photojournalism collective noting that 'many of the best known names in photojournalism have photographed the ship-breaking yards in Chittagong.' It is

Figure 4.2   Hidden worlds of wasting large enough to be seen from space
Source: © 2010 Google, © 2010 DigitalGlobe, © 2010 Mapabc.com

thus a site where the different registers through which materiality is pictured can become visible. The focus on temporalities of decline and materiality are common yet distinct if we follow two of the most celebrated photographers to have looked at this site – Sebastiao Salgadao and Edward Burtynsky.

Tropics out of Time

Perhaps the first photographer to depict the breaking beaches of Chittagong was Sebastiao Salgado, whose work also fronts one of the exposés of labour conditions on cover of *Atlantic* magazine and forms a section in his collection *Workers: An Archaeology of the Industrial Age* (2005) that documents the current state of manual labour around the globe. His work evinces a preoccupation with enabling workers to become visible but also to return the gaze, sometimes literally, of the consumers. His account of the ship breaking beaches of Bangladesh breaking stresses the agency and action of the workers who:

> run these ships onto the beach at high speed; then they attack them from all sides, blow torches cut through its steelskin, giant hammers break up its iron and wood structure ... Everything from that giant animal lying on the beach has its use. Iron and steel will be melted down and given new roles as utensils. The entire ship will be turned into what it once carried: machines, knives and forks, hoes, shovels, screws, things, bits, pieces ... The huge bronze propellers will provide the most elegant of items – bracelets earrings, necklaces, and rings which will one day adorn the bodies of working women, as well as pots from which men will pour tea (Salgado, 2005: 14).

Steel ship becomes dying animal. It is disassembled into hand sized objects. The description carries the faint whiff of exoticism, where workers in the global south seem always to throng, and the objects which the ship becomes make sure to mention tea and bracelets but not the rather more modern reinforcing rods in ferro-concrete houses which will be the fate of 90 per cent of the ship's steel (Figure 4.3). Neither the artisanal reworking of materials, nor the home furnishings that are produced, though, capture the lingering gaze of the photographer. The rebirth and recycling of steel do not fit the narrative arc of these pictures. Salgado reworks twentieth-century progressive-era photography narratives about the accomplishments of labour building vast edifices. Salgado takes the modernist style and inverts its narrative. So where modernism developed an industrial sublime that 'used to glorify leviathan industry and occasionally its workers, it is here turned to elegy' and we have 'masked workers, and fragile figures pressed up against gigantic machinery; ... all recast by Salgado in a period of decline' (Stallabrass, 1997: 149). In part his pictures are an assertion of these worlds of manual toil in an age that often claims to be post-industrial; they are images out of time that echo the worst labour conditions that many in the west have consigned

**Figure 4.3   The afterlives of steel: ships cable becomes reinforcing rod for
buildings, The Waste of the World**

to history. He offers a lingering 'farewell to a world of manual labor that is slowly
disappearing' from the consciousness of the west yet persists out of sight, and
concentrates on the antiquated heroism of physical toil (Ziser and Sze, 2009: 397).
Indeed especially in the ship breaking pictures the sense of decline is manifest in
'dramatic images of strenuous labour, of mighty hammer blows struck and great
weights borne, not to make some grand vessel, but to take one apart' (Stallabrass,
1997: 150) (Figure 4.3).

The fragility of human life appears in poignantly small touches in Salagado's
pictures – where workers toil across the devastated beach to bring a cable out
to winch in part of a ship, their exposure to the materiality of the ships and the
shore is emphasized by the diminutive umbrella the supervisor uses as the sole
protection from the sun. Stallabrass argues here we have an invocation of religious
imagery where:

> part of the point about the pictures of workers before fire, or fumes or giant
> machinery [is that] these involuntary neophytes are sacrificed to the numerous
> deities – commodities and corporations – of the capitalist cosmos. In this sense,
> to present workers as battling against forces beyond their control is to tell an
> uncomfortable truth. While humanity may be one, the gods, like those of the
> ancient world, are many and warring, and reckless in the use of their human
> charges (page 152).

The human figures are oppressed and outscaled by the huge materials – in visions where we see not progress or technological triumph, but capitalism as disaster.

Entropic Ruins

Edward Burtynsky's work on industrially altered landscapes for the last thirty years offers a richer sense of the material recomposition of wastes. Over that period his corpus of work extols the material counterpoint of urban and industrialization, looking at mines and quarries as the negative image of cities and factories. His work focuses upon the sources and destinations of the material flows of an increasingly globalized industrialization. In this sense he works with what he calls the residual landscape of the consequences of capitalism. His recent work has been gradually picking out different points in the global circulation of commodities, such that he offers 'a chronicle of rarely seen points in the biography of everyday things and work environments ... [p]resented as a kind of material culture of globalization' (Campbell, 2008: 40). He maintains a studied ambiguity, aesthetic and ideological, about the epic scale and grandeur of industry and sites of extraction as an industrial sublime. He focuses upon the materials, not the labour, so in his shipbreaking pictures the workers 'seem almost allegorical, ant like men fragmenting the colossi that are the only relief in that vast, flat expanse' (Solnit, 2007: 138). The composition has a democratic distribution of light and space (Burtynsky, 2003: 52; Baker, 2003), that moves it away from an anthropocentric perspective. The pictures often disorient senses of scale and dislocate any human focal point – as subject or viewer. These pictures foreground material transformation as a necessary and ongoing part of our world, and thus retain an attachment to the sensuality of the world (Solnit, 2007: 139).

Burtynsky thus creates 'a space of witnessing that is not coincident with a particular form of [political] agency' (Zehle, 2008: 113). While documenting the unseen world may be political (Solnit 2007: 137), the mode of witnessing is rather different here. The pictures focus less upon indexical truth claims (that this happened there), but become iconic symbols (for the process at large).

> To me, what is interesting as an artist, or mediator, is to reconnect to the sources of our lifestyle, to find a way to capture the immensity of scale and activity there, but not in what most think of as a purely 'documentary' fashion [about a specific example but rather choosing one] that somehow has a special quality that allows me as a photographer to transform into something that goes well beyond the thing itself (Burtynsky cited in Campbell, 2008: 42).

The material referent's hold on the pictures is weakened in order to enhance the aesthetic quality (Cammaer, 2009: 122). Burtynsky uses form to connect with the larger and the abstract, both quantitatively and conceptually. This is partly done by reproducing the pictures at large scale. So his work is marked out by

Figure 4.4　　The labour of reduction, The Waste of the World

use of large format cameras and large format pictures with static and formally composed subjects. This affects his photographic practice and vision, as he uses a:

> 8×10 large-format 'bellows' camera … Burtynsky has to use a tripod to stabilize the camera; a black focusing cloth is also necessary (to darken the image reflected on the camera's exposed ground glass plate while the practitioner adjusts focus). The symbolic logic of this artistic practice – the artist stooped beneath the black shroud, finger poised on the cable release button, faithful to an old-fashioned mode of photography – suggests a defiance of the reckless forward momentum of industrialization (Bozak, 2008/2009: 71).

Indeed his recent exhibition of field proofs – polaroid scoping shots, blown up large scale, but in black and white marked with the creases and tears of their use – serves both to restate the presence of the photographer, and the materiality of the practice, whilst simultaneously making the pictures appear more anachronistic. The pictures call to mind 'the expeditionary images of ancient monuments in exotic lands taken by nineteenth century photographers' (Baker, 2003: 51). In his Chittagong #8, as Diehl (2006: 121) suggests, 'the massive odd-shaped sculptural elements in this play of light and shadow suggest the ruins of an ancient metropolis unknown to Westerners.' The temporality we might look to is that of the ruin. In this his pictures evoke a lost order, perhaps echoing Thomas Cole's famous conclusion of his five-part *The Course of Empire* (1836), which charts a cycle of rise and fall from 'savage state' to a 'pastoral' to the pinnacle of urban civilization and then to 'destruction' and finally 'desolation,' full of the ruins that symbolize 'a culture on the cusp of disappearance at the hands of natural and man-made forces' (Ziser and Sze, 2009: 395). Burtynksy himself has said he is interested in the 'ruins of our society' as both melancholy and monumental (Zehle, 2008: 111). Notably then we do not get 'destruction' as process despite the echoes of visions of industrial hell from paintings such as Philip James de Loutherbourg's *Coalbrookdale by Night* (1801) that are referents in other photographers' work which also play on the flames and smoke of cutting.

The sublime is evoked as much as the melancholy through the scale of the images – not only by speaking to global space and large scale pictures but also by freezing and yet revealing time. In terms of content some have argued that 'Burtynsky's photographs are of time sublime – an empty now and a terrifying prospect' (Giblett, 2009: 785–6). Burtynsky has an elegiac sense so that his:

> 'Shipbreaking' photographs, like all his works, appear to us as images of the end of time. The abandoned mines and quarries, the piles of discarded tires, the endless fields of oil derricks, the huge monoliths of retired tankers show how our attempts at industrial 'progress' often leave a residue of destruction. Nevertheless there is something uncannily beautiful and breath taking in the very expansiveness of these images – it is as if the vastness of their perspective somehow opens onto the longer view of things (Pauli, 2003b: 33).

And then we must add the moment of audiencing for these picture. Burtynsky produces pictures with a painterly effect that slows viewing from the snapshot to contemplation (Burtynsky, 2003: 48).

Materialisms, Metal and Waste

On the beaches of Chittagong, then, we find photographic approaches that in part play upon the documentary function of exposing the distant and hidden. They contain a revelatory charge even if the open beach is large enough to be seen from space. The different modes of accessing the lives of the beach speak to different visualizations. Our remote sensing of the site is rather too remote and too denuded of senses to convey the material conditions. Accessing the site up close, as the pictures used here do, requires negotiating the walls and security fences in order to show the actions and activities rather than the traces left visible from space. Both Salgadao and Burtynsky offer pictures that speak to grand changes and processes in political economy and ecology. The pictures of the Chittagong foreshore also address different registers of materiality. In this sense then the scale of Burtynsky's work offers literally a big picture:

These big pictures are attempts to map *the* big picture, to render visible those zones where power moves and possibilities are both generated and shut down … It is one thing to read about the scale of contemporary factories and to learn from reportage of sheer numbers about the volume of money sloshing around the world on a daily basis. It is another thing entirely to see it rendered visible – a visibility which is not documentary in an immediate and unproblematic sense, but which generates knowledge of a kind that only an image can manage to do … The power of contemporary photography derives not just from the first-order operations of visualisation, but from its unique aesthetic and political motility – its ability to both use and refuse older aesthetic categories and determinations in conjunction with the mechanics of the photo apparatus and the digital flow of networks to provide conceptual maps we would not otherwise have (Szeman and Whiteman, 2009: 554).

Both offer pictures that speak to time lost and places depleted. Salgado depicts work whose persistence is the dirty secret of globalization. In his urgency to have workers return our gaze, he ends up telling a story of material culture around the ships – they are things bestowed biographies through human agency (Bennett 2010). Burtynsky's images instead attend to the actual material decomposition of the ships on the beaches of Chittagong, and the very substance of the wastes.

Figure 4.5   A parade of ships becoming steel, The Waste of the World

These pictures all offer a way of apprehending – or more specifically apprehending through images – something of the awesome and awful power of capitalist globalization. It is intriguing that they seem by offering a frozen moment to open out the material times, spaces and consequences of global capitalism. The capacities and audiencing of these pictures enables them to speak in synch with a particular set of transformations. For Salgado it is the passage, persistence and witnessing of manual labour, for Burtynsky it is the dynamic transformation of materials. In the first, it is a hylomorphic world where material is animated by humanity; in the latter humanity is displaced. The question is how far? Although the steel lives – in its metallic structures and crystalline properties – the ships do not simply become ruins as happenstance wrecks on the beach. The labour and work of destruction is massive – and also directly connected with the messy material afterlives of the steel. The steel is hacked from the hulls, to be dragged across mudflats by people all too exposed to the materiality of the steel and its coatings, and then rerolled in manual mills where it is reborn as reinforcing rods – and yes then its microscopic crystalline life is once more in play as the ductile qualities of ship steel are most amenable to that simplest, and cheapest, form of metal refabrication.

For Salgado, and many other more photojournalistic approaches to the site, the materiality that matters is the impact of ships and their breaking upon human subjects. The dangerous, damaging labour in tropical heat is a register of embodied materiality that is salutary and important to capture. Yet still the mode of practice and form of vision seem somehow atavistic and out of place – like the static tableaux of colonial typologies. In part, this is a deliberate restaging that speaks the persistence of that imaginary and those patterns of power. In part though there also seems to be an elegiac melancholy for lives lost and things destroyed. The charge is of toxic exotique, of the vibrant and verdant become death. Indeed a tropology of toxicity is set against the imagined otherness of the tropical locale. To engage the material life of steel offers a different charge. Here the register remains one of entropy and dissipation, not one of rebirth. Yet the breaking of ships is founded on the capacity of steel for almost endless recycling, refashioning and reuse. Still photography though figures the nonorganic life of steel in elegiac mode. It is as if it too had looked upon the changing world and with Claude Levi-Strauss (1961: 397) concluded that the history of humanity is one of 'cheerfully dismantl[ing] million upon million of structures and reduc[ing] their elements to a state in which they can no longer be reintegrated' and through destruction flattening difference – be it humans and/or materials – until global change requires less a discipline of anthropology than entropology to study it.

Figure 4.6   Toxic tropics: tropes of exotic toxicity, The Waste of the World

References

Baker, K. (2003) Form versus portent: Edward Burtynsky's engandgered landscapes. In: L. Pauli (ed.) *Manufactured Landscapes: The Photographs of Edward Burtynsky* (New Haven: Yale University Press, CT, 40–45).

Benjamin, W. (1978) *A Small History of Photography One Way Street & other writings* (London: Verso, 240–257).

Bennett, J. (2001) *The Enchantment of Modern Life: Attachments, Crossings and Ethics* (Princeton: University Press Princeton).

Bennett, J. (2010) *Vibrant Matter: A Political Ecology of Things* (Durham: Duke University Press).

Bozak, N. (2008/2009) Manufactured landscapes. *Film Quarterly*, 62, 68–71.

Burtynsky, E. (2003) *Manufactured Landscapes* (New Haven: Yale University Press).

Cammaer, G. (2009) Edward Burtynsky's manufactured landscapes: the ethics and aesthetics of creating moving still images and stilling moving images of ecological disasters. *Environmental Communication: A Journal of Nature and Culture*, 3:1, 121–130.

Campbell, C. (2008) Residual landscapes and the everyday: an interview with Edward Burtynsky. *Space and Culture*, 11:1, 39–50.

Chandler, D. (2000) Drift and fall. In: J. Southam (ed.) *The Shape of Time: Rockfalls, Rivermouths, Ponds* (London: Photoworks).

de Landa, M. (1992) Nonorganic life. In: J. Crary and S. Kwinter (eds) *Zone 6: Incorporations* (New York: Urzone, 129–167).

de Landa, M. (2006) Matter matters. *Domus 888*, January 2006, 136–37.

Diehl, C. (2006) The Toxic Sublime. *Art Am*, 94:2, 118–123.

Giblett, R. (2009) Terrifying prospects and resources of hope: minescapes, timescapes and the aesthetics of the future. *Continuum: Journal of Media & Cultural Studies*, 23:6, 781–789.

Jay, M. (1993) *With Downcast Eyes: The Denigration of Vision in Twentieth Century French Thought* (Berkeley: California University Press).

Lévi-Strauss, C. (1961) *Tristes Tropiques*, J. Russell (trans.) (New York: Criterion Books).

Maskit, J. (2007) Line of wreckage: towards a postindustrial environmental aesthetics. *Ethics, Place & Environment*, 10:3, 323–337.

Miles, M. (2002) Seeing through place: local approaches to global problems. In: J. Rugg and D. Hinchliffe (eds) *Recoveries and Reclamations* (London: Intellect Books, 77–90).

Millar, J. (2000) A gaze equal to space. In: J. Southam (ed.) *The Shape of Time: Rockfalls, Rivermouths, Ponds* (London: Photoworks).

National Labor Committee (2009) *Where Ships and Workers Go to Die: Shipbreaking in Bangladesh and the Failure of Global Institutions to Protect Workers' Rights*, National Labor Committee.

Olin, M. (2002) Touching photographs: Roland Barthes's 'mistaken' identification, *Representations*, 80, 99–118.

Pauli, L. (ed.) (2003a) *Manufactured Landscapes: The Photographs of Edward Burtynsky* (New Haven: Yale University Press).

Pauli, L. (2003b) Seeing the big picture. In: L. Pauli (ed.) *Manufactured Landscapes: The Photographs of Edward Burtynsky* (New Haven: Yale University Press, 10–33).

Rugg, J. (2000) Regeneration or reparation: death, loss and absence in Anya Gallacio's intensities and surfaces and forest floor. In: S. Bennett and J. Butler (eds) *Locality, Regeneration and Divers(c)ities: Advances in Art and Urban Futures*, Volume 1 (London: Intellect Books, 41–50).

Rugg, J. (2007) Fear and flowers in Anya Gallaccio's forest floor, keep off the grass, glaschu and repens. In: F. Becket and T. Gifford (eds) *Culture, Creativity and Environment: New Environmentalist Criticism 5* (Amsterdam: Rodopi, 55–74).

Russell, B. (1918) *Mysticism and Logic and Other Essays* (London: George Allen & Unwin).

Salgado, S. (2005) *Workers: An Archaeology of the Industrial Age* (New York: Aperture).

Serres, M. and Hallward, P. (2003) The science of relations. *Angelaki: Journal of Theoretical Humanities*, 8:2, 227–238.

Smithson, R. (1996) *The Collected Writings* (Berkeley: University of California Press).

Solnit, R. (2007) *Storming the Gates of Paradise: Landscapes for Politics* (Berkeley: University of California Press).

Sontag, S. (1977) *On Photography* (London: Penguin).

Southam, J. (2000) *The Shape of Time: Rockfalls, Rivermouths, Ponds* (London: Photoworks).

Stallabrass, J. (1997) Sebastião Salgado and fine art photojournalism. *NLR 223*, 131–161.

Szeman, I. and Whiteman, M. (2009) The big picture: on the politics of contemporary photography. *Third Text*, 23:2, 551–556.

Webb, D. (2006) Michel Serres on Lucretius. *Angelaki*, 11:3, 125–136.

Wightman, J. (2008) Winogradsky rothko: bacterial ecosystem as pastoral landscape. *Journal of Visual Culture*, 7:3, 309–334.

Yusoff, K. and Gabrys, J. (2006) Time lapses: Robert Smithson' mobile landscapes. *Cultural Geographies*, 13:3, 444–450.

Zehle, S. (2008) Dispatches from the depletion zone: Edward Burtynsky and the documentary sublime. *Media International Australia*, 127:May, 109–115.

Ziser, M. and Sze, J. (2009) Climate change, environmental aesthetics, and global environmental justice cultural studies. *Discourse*, 29:2–3, 384–410.

Acknowledgements

The work here is from the project 'The Waste of the World' funded by the Economic and Social Research Council (RES-060-23-0007).

Chapter 5
Citizen and Denizen Space: If Walls Could Speak

Nirmal Puwar

It is a delimination of spaces and times, of the visible and the invisible, of speech and noise, that simultaneously determines the place and the stakes of politics as a form of experience. Politics revolves around what is seen and what can be said about it, around who has the ability to see and the talent to speak, around the properties of spaces and the possibilities of time.

Ranciere, 2004: 13

Walls are an underlying refrain in this chapter. Walls, cities, states and art are relational assemblages (Shanks, 1999; Wiezman, 2007). In *If Walls Could Speak/ Si Las Paredes Hablaran* (1991), the artist Celia Alvarez Munoz, working on the union history of Mexican workers in Los Angelos through the activities centred in the Embassy Building, states "If walls could speak, these walls would tell/in sounds of music voices, music and machines/of the early tremors of the City of Angels" (cited in Hayden, 1997: 201). For some time it has been recognized how history, as well as memory, are place based (Casey, 1998; Nora, 1992). Buildings are palimpsests; of what we sense as being there before, as well as alternatives to what is there now (Huyssen, 2003: 7). I turn to the walls of parliaments in order to work with the constitutive boundaries of nation making. The archi-textures of both stable and tenuous demarcations are considered – beyond any materiality/visuality dichotomy – both inside/outside the walls, as well as the perimeters. This chapter is presented as a textured collage. It simultaneously *performs* a critical reflection on what is sayable and visible in the constitution of citizen/denizen space and puts forward imaginary re-calibrations of parliamentary sites.

Composed of three parts, the chapter considers the visuality of materiality as manifested in the contours of (parliamentary) citizen and denizen space(s). Treated as embodied practices, the very textures of how spaces are produced, demarcated, sedimented, disrupted and re-invented, assumes the imbrication of materiality with visuality. Amidst the vexed theoretical debates of the properties and weight assigned to language, object, biology, image, the relational, the affective and the body, there have been a number of "turns". This article is no doubt informed by a number of debates couched within the material turn found in anthropology (Miller, 1987), visual practices (Rose, 2001) and actor-network-theory (Latour, 2005a), biography and movement (Appadurai, 1998; Thrift, 2007) without actually being a direct adherent of or homage to any of these theoretical perspectives. The key terms of the argument developed are: texture, politics and space. The rubric of texture is understood as a dynamic and sedimented inter-relational approach to objects (visual or architectural), bodies and power (cf. Bruno, 2002; Stoller, 1997).

Walking the Walls

Rhesus macaques monkeys walk the ledges of the Indian Parliament in New Delhi. They court the gardens and fountains of the estate. Within the imperial plan of governance, the British planned New Delhi as the capital of the country, moving it from Calcutta in 1911. Designed by the architects Edwin Lutyen and Herbert Baker, Parliament House was opened at a ceremony performed by the then Governor-General of India, Lord Irwin, on 18th January 1927. This is a huge circular structure, five-hundred and sixty feet in diameter and one third of a mile in circumference. In a special issue on "New Delhi", of the journal *Architectural Review* (1931), Robert Byron described the chamber thus: "It resembles a Spanish bull-ring, lying like a mill-wheel dropped accidentally on its side."

The imprint of empire is sculptured into the stone and engravings. The columns at the front entrance of the palace have bells carved into them which, it has been suggested, Lutyens had designed with the idea that while the bells were silent British rule would never come to an end. The wood-work in the Lok Sabha Chamber has 35 gilded designs representing the various Provinces of undivided India, the Dominions and certain other British Settlements. While the design is not neo-gothic as were other buildings which were designed by the British in India and inspired by Westminster – such as the town hall in Bombay, the Gateway to India and the Victoria Terminus – the colours of the elected (green) and the unelected (red) members of Parliament are replicated. Needless to say, the European presence looms large in the hybrid mixture of Indo-Saracenic architecture.

Today, all too frequently, the twelve iron gates of Parliament House provide grab rails for monkeys. Ripping cabinet documents, ransacking secret files, they swiftly move in groups through the iron grills that enclose the Parliament House estate, wrapping themselves around the architecture of the buildings, hunting and hooting from one quarter to another. As if the in-tray of the Defense ministry was not already over flowing with complaints of malpractice, the monkeys have even managed to get their hands on secret official documents. Even ultrahigh frequency loud speakers could not deter the monkeys who hung out from window ledges screeching at reporters when they tried to hold a press conference with the visiting US Secretary of State Donald Rumsfeld in 2004. Not being able to secure a gated community without the disturbing presence of these animals, Languar primates have been bought in as private security to patrol the grounds. Walked on a leash, their owners make a living by using the Languars to scare off the Rhesus monkeys, who in turn move on to another patch in the large government compounds. While official visitor signs on the grounds state "Please do not feed the monkeys", feeding continues. Rather ironically, the (monkey god) Hanuman temple sits in close vicinity, on Connaught Place, where rituals ceremoniously worship the monkey for being a loyal and intelligent servant, who in the Hindu epic Ramayana, led an army of monkeys to fight the demon king Ravana. Because the monkeys stalk the parliamentary compound (consisting of offices as well as residences), calls for a monkey-free city have been loudly proclaimed in the court rooms. Attempts to deport monkeys to neighboring states and detention prisons have been made. Contesting these environmentalists, including Maneka Gandhi (Indian politician and environmentalist, who is herself a repelled member of the Gandhi dynasty), point to increased deforestation in the surrounding areas which has led monkeys to go in search of inhabitations and food within the drawers and cabinets of government staff and ministers. Who are the citizens and denizens!

Space is not only, as pointed out in Henri Lefebvre's *The Production of Space* (1991) "abstract", that is subject to calculus and homogenous control. But it is also "lived", as an inter-woven series of local encounters, involving sensuous connections and imagination. Parliamentary spaces are thus party to the spatial engagements of MPs, journalists, publics, service staff and even monkeys, as well as others deemed denizens and not citizens. Auditory rhythms of communication within and beyond the walls distinguishes what Ranciere calls the "sayable" and visible in the "distribution of the sensible" (2004: 41). While some voices fill the architectural volume of the buildings with speech, that is both spoken and heard (Spivak, 1988). Others are assigned the status of the "hysterical" (Gatens, 1996: 54) and "noise" which is chaotic, wild and disruptive (Attali, 1985) or "noise" as turbulence and nuisance (Serres, 2007). Never the less outsider and denizen bodies continue to push at the limits – on different grounds – of what constitutes the

Figure 5.1   Monkeys
Source: Photo by Kelly Nicoll

political citizen (Pateman and Mills, 2007; Puwar, 2004). Oppositional voices can challenge the politico-aesthetics of the somatic human norm of politics through perhaps an instigation of "sensate democracy" (Butler, 2006: 15; 2007: 62).

The Nation on the Wall

The palimpsest nature (Huyssen, 2003) of the specific site of the British Parliament has sensed several lives; palace, law court, church, debating chamber and club. With these changes the commons has also moved locations in the building. Shifting powers have in time added new buildings and inventions, just as fires and war bombings have destroyed. Each time it has been re-built there has been controversy over the style, the shape and the size. Which heritage should it follow and which boundaries should it produce anew? Whose body should it be molded for? Which images and sounds should fill it? The oldest building in Parliament is Westminster Hall. If walls could speak its lives, we would hear coronation banquets, the feasting of palaces, courts and traders selling wigs and scribes. None of these materialities have totally left Parliament. Yes they have been re-ordered and re-cast. But their presence – of religion, monarch and empire – is still more than a trace. The haptic qualities of Westminster are plenty but to what purposes are the visuality of this materiality chanelled. Politics is conducted with texture, performance, furniture, cloth, sound and bodies in space. The so called writing on the walls (and even the floors) can, in Paul Connerton's analysis of how nations memorialize through the writing of history, be understood as, "the formation of a political identity and giving shape to the memory of a particular culture" (1989: 16).

The walls of the ninety-five feet long vaulted St Stephen's Hall include large mural paintings commissioned by Henry John Newbolt under the theme of *The Building of Britain*, completed in 1927 (Newbolt, 1927; and see the virtual tour at <http://www.parliament.uk/visiting/online-tours/virtualtours/ststephenshall_ and_chapel_cloister/>). The murals chart battles, a selection of "key moments" and expeditions of what are understood to be significant markers in the birth and growth of Britain as a nation and empire across the Commonwealth. King Alfred's ships built to defend against Danish invasion, King Richard I setting off for battle in Jerusalem, King John unwillingly consenting to Magna Carta, the tension between King and the Commons, the parliamentary union, the Queen blessing Sir Walter Raleigh to discover new lands in the Americas, as well as Thomas Roe from the East India Company accessioning a commercial treaty in the court of Ajmir in 1614, are all visually displayed on the walls of the hall. Commenting on *The Building of Britain* paintings, Claire Willsdon has argued that, despite parliamentary disagreements on the content of the murals, what went up was concerned with "conciliation, consensus and pastoral idealism in the maintenance of Imperial stability" (2000: 139).

Having mutated from chapel, to commons chamber and now a grand hallway, St Stephen's Hall has been through changing uses, proportions and built materials, not least because of the Great Fire of 1834, which reduced it to ruins. Hence the influence of the architect Charles Barry, who was granted the commission in 1835, and the hand of Augustus Welby Pugin, prevails in the design. Light moves in through ten stained-glass windows, five on either side, depicting the arms of various parliamentary cities and boroughs. Statesmen and Norman and Plantagenet monarchs stand, as stone ghosts, proudly tall, over those who walk through the hall today. Two mosaics of the founding of the chapel by King Stephen and its rebuilding by Edward III also feature. Both figures are stood at the side of St Stephen holding a stone, in allusion to his martyrdom. Interestingly, in one of the few visible references to the hidden labour which has built Westminster during the course of its different lives, a mosaic at the west end, unveiled in 1926, depicts Edward III approving the plans for the chapel and handing them back to his master mason, Michael of Canterbury, with representatives of medieval craftsmen standing beside him. Leading on from the hall to the octagonal central lobby, here, over each of the four exits from the lobby, are large mosaic panels, of St. George, St. David, St. Andrew and St. Patrick.

Figure 5.2   St Stephen's Hall, Westminster
Source: © Parliamentary

What is not visible, in terms of the history of St Stephen's Hall, is the stencil graffiti stamped on the wall by the suffragette, sculptor and illustrator Marion Wallace-Dunlop. On 22nd June 1909 she used the wording from the 1689 Bill of Rights, which had been ratified in the very hall itself. Originally, the Bill stated: "It is the right of the subjects to petition the King and all commitments and prosecutions for such petitioning are illegal". Using violet indelible printer's ink, she twisted it around capitalizing "Subjects" but not "king". The same words had been used by the suffragettes on leaflets, chalked pavements and delivered in speeches. Using typography to disturb what was "sayable" to alter what is visible, thinkable and possible. Thus, challenging both the aesthetic and gendered boundaries of politics. Caught in mid-action, the guards confiscated her stamp and ejected her from the hall. She returned two days later, tooled up with a second stencil cut from a sheet of rubber. On this occasion she managed to successfully etch text but was arrested and imprisoned. From Holloway Prison in London she continued politics through her body, rather than her art and typography, by famously publicizing the hunger strike as a political weapon (Puwar, 2010).

Blood on the Walls

Just as violence is performed spectacularly in the aggressive (pantomime) jaunts of parliamentary members in chamber debates, violence is also mediated by the bureaucracies of democracy, globally and locally.

In 2005 the Serious Organised Crime and Police Act (SOCPA) restricted the right to demonstrate within an exclusion zone of up to 1 kilometre around Parliament. The changes meant that demonstrators have to apply for permission to the Metropolitan Commissioner six days in advance (or no less than twenty-four hours if six days are not practicable). Conditions can at any time be attached to the demonstration. Loudspeakers are banned except for use by those in positions of authority. The legislation was initiated in response to Brian Haw who had demonstrated in Parliament Square, since 2002, as a peace activist against the American and British invasion of Iraq, using placards and loudspeakers. The first conviction under the Act was in December 2005, when Maya Evans read the names of British soldiers killed in the Iraq War, near the cenotaph without police authorization (see Evans and Rai, 2006).

The comedian Mark Thomas targeted the act by organising twenty-one protests across a single day in the designated area within the confines of the law. Brian Haw has managed to remain on the site even though his items were bulldozed and removed from the site by the police on 23rd May 2006. In reaction, the artist Mark Wallinger recreated Brian Haw's display which was confiscated by the police for an exhibition titled *State Britain* in Tate Britain, for which he won the Turner Prize (2007). Over 600 items were remade, found and assembled. They included banners, placards, posters, peace flags, newspaper articles, photo displays, messages from supporters and teddy bears wearing peace-slogan t-shirts.

Figure 5.3   **Mark Wallinger, *State Britain*, 2007, detail. Mixed media installation at Tate Britain, London, 570cm × 190cm × 43m**

Source: Photo by Dave Morgan; © the artist; courtesy Anthony Reynolds Gallery, London

The effects of the weather and the environment on the items were also recreated in materializing the installation. Wallinger marked a black line on the floor to signal the area of the installation inside the exclusion zone, running through the Tate. The press even speculated whether the police would remove parts of the exhibit which were in the kilometre radius zone around Parliament.

Due to continued complaints and issues, the government has agreed to abolish the legal requirements for protests around Parliament to be notified in advance to the police. However, GLA's byelaws remain restrictive in terms of prior permission (from the Lord Mayor) as well as the use of banners, placards or any sound amplification device.

Meanwhile art installations on war continue to grow at demonstrations as well as in galleries. For instance, having seen billboards marketing the exhibition all over the city, including the London underground and public phone booths, paying public carried themselves to the Royal Academy (2009) on Piccadilly to Anish Kapoor's installations. To great saturated effect the publicity posters, seen all over the city, featured the red blood (wax) of the work which was the highlight of the exhibition. *Shooting into the Corner* was an installation drama. Red wax was smacked against a corner of the room. The red wax builds up against the white walls and floor of the gallery. The audience lay in wait of the splattering that took place every twenty minutes when an attendant picked up a pre-prepared weight of red wax, carefully slotted it in the canon, before shooting it into the corner. Observance of the loading of the canon by the audience was central to the installation. Some audience members, which included large numbers of tourists, covered their ears with their hands, in wait of a loud bang on the shoot. Thus part performance, part fun fare and part respectable art making and viewing within the consecrated walls of the British Academy.

Perhaps an even more apt venue today could be Westminster Hall, of which the walls were built in 1097–99 and where many a bloodshed has been instituted and enacted. Indeed on 30th January 1661, marking the restoration of the monarchy, as well as vengeance for the execution of Charles I, Charles II had Oliver Cromwell exhumed from Westminster Abbey and subject to a posthumous ceremony of execution. Cromwell's head was stuck on a post and displayed in Westminster Hall. Only a storm blew it down some twenty-years later. Westminster Hall is used today for major public ceremonial events such as lyings in state. There is much to be said of the sensory and the haptic in the visual and material machinations of this building, but a turn to an appreciation of the visceral and texture (as we have seen in recent moves towards the assertion of the study of the senses), does not simply awaken touchy feely encounters of the non-violent kind. Let it take Kapoor's canons of red wax, that pile up to resemble the build up of blooded horse hooves left behind on a macabre battlefield, with unknown human body parts blanketed by the dense texture of the many bloods that have been spilt. Could this even be an

**Figure 5.4   Anish Kapoor, *Shooting into the Corner*, 2008–2009. Mixed
             media installation at Royal Academy of Arts, London,
             dimensions variable**
Source: Photo: Dave Morgan; courtesy: the artist

alchemy for a hybrid exhibitionary/parliamentary "phantom public", as postulated by Bruno Latour (2005), in relation to the *res publica* of exhibition spaces?

References

Appadurai, A. (1998) *Modernity at Large: Cultural Dimensions of Globalization* (Minneapolis: University of Minnesota Press).

Attali, J. (1985) *Noise: The Political Economy of Music* (Minneapolis: University of Minnesota Press).

Butler, J. (2006) *Precarious Life: The Powers of Mourning and Violence* (London: Verso).

Butler, J. (2007) *Spivak Gayatri Chakravorty: Who Sings the Nation-State?: Language, Politics, Belonging* (Oxford: Seagull Books).

Byron, R. (1931) New Delhi, *Architectural Review*, LXIX.

Casey, E. (1998) *The Fate of Place* (Berkeley: University of California Press).

Connerton, P. (1989) *How Societies Remember* (Cambridge: University of Cambridge Press).

Evans, M. and Rai, M. (2006) *Naming the Dead: A Serious Crime* (Drava Papers).

Gatens, M. (1996) *Imaginary Bodies: Ethics, Power and Corporeality* (London: Routledge).

Giuliana, B. (2002) *Atlas of Emotion: Journeys in Art, Architecture and Film* (New York: Verso).

Grosz, E. (2001) *Architecture From the Outside: Essays on Virtual & Real Space* (Cambridge: MIT Press).

Hayden, D. (1997) *The Power of Place: Urban landscapes as Public History* (Cambridge: MIT Press).

Huyssen, A. (2003) *Present Pasts Present Pasts Urban Palimpsests and the Politics of Memory* (Stanford: Stanford University Press).

Latour, B. (2005a). *Reassembling the Social: An Introduction to Actor-Network-Theory* (Oxford: Oxford University Press).

Latour, B. and Weibel, P. (eds) (2005b) *Making Things Public – Atmospheres of Democracy* (Cambridge: MIT Press).

Lefebvre, H. (1991) *The Production of Space* (Oxford: Blackwell).

Miller, D. (1987) *Material Culture and Mass Consumption* (Oxford: Basil Blackwell).

Munoz, C.A. (1991) *If Walls Could Speak/Si Las Paredes Hablaran* (Arlington: Enlightenment Press).

Newbolt, H.J. (1927) *The Building of Britain* (T. Nelson & Sons)

Nora, P. (1992) From lieux de to realms of memory in *Realms of Memory; Rethinking the French Past.* (ed.) P. Nora (New York: Columbia University Press, 1–23).

Nunn, H. (2002) *Thatcher, Politics and Fantasy* (London: Lawrence and Wishart).

Pateman, C. (1995) *The Disorder of Women: Democracy, Feminism and Political Theory* (Cambridge: Polity Press).

Pateman, C. and Mills, C. (2007) *Contract and Domination* (Cambridge: Polity Press).

Puwar, N. (2004) *Space Invaders: Race, Gender and Bodies out of Place* (Oxford: Berg Publishers).

Puwar, N. (2010) The archi-texture of parliament: flaneur as method in Westminster. *The Journal of Legislative Studies*, 16:3, pp 298–312).

Ranciere, J. (2007) *The Politics of Aesthetics* (London: Continuum).

Rose, G. (2001) *Visual Methodologies: An Introduction to the Interpretation of Visual Materials* (London: Sage).

Serres, M. (2007) *The Parasite* (Minneapolis: University of Minnesota Press).

Shanks, M. (1999) *Art and the Early Greek State: An Interpretative Archeaology* (Cambridge: Cambridge University Press).

Spivak, G.C. (1988) Can the subaltern speak? in *Marxism and the Interpretation of Culture*. (eds) C. Nelson and L. Grossberg (Urbana: University of Illinois Press, 271–313).

Steedman, C. (2005) Poetical maids and cooks who wrote. *Eighteenth-Century Studies*, 39:1, pp 1–27.

Stoller, P. (1997) *Sensuous Scholarship* (Philadelphia: University of Pennsylvania Press).

Thrift, N. (2007) *Non-Representational Theory* (London: Routledge).

Trüby, S. (2008) *Exit-architecture: Design between War and Peace* (New York: Springer Wein).

Weizman, E. (2007) *Hollow Land: Israel's Architecture of Occupation* (London and New York: Verso).

Willsdon, C.A.P. (2000) *Mural Painting in Britain 1840–1940: Image and Meaning* (Oxford: Oxford University Press).

Chapter 6

Seeing Air

Caren Yglesias

People in the design fields acknowledge their art is made by others. What designers *do* make are representational drawings envisioning prospective constructions, both to explore design ideas and to build those ideas. The question central to this discussion is in what way does visual thinking benefit from drawing, not just abstract forms, but also materials? Further, can such drawings cultivate a designer's sense of projective affect thereby enhancing the design of places and projects? The first part of this chapter presents a method of exercising the material imagination using the homophones: *sighting*, how perception is cultivated by training the eye to look thoughtfully; *siting*, where critical relationships between object and setting are recognized; and *citing*, when external referents are summoned to enhance depth of understanding. This exercise prepares the visual imagination to make drawings using techniques of material representation. These proposed techniques loosely follow the twentieth-century philosopher R. G. Collingwood's aesthetic categories, an interpretation of which suggests a sequence of four drawing types. This is the focus of the second part of this chapter. In other words, I argue that architecture and landscape architecture – the useful or productive arts – develop by drawing physical materials materially. This results in a greater capacity to link intention with affective experiential response. Using Cézanne as an illustrative example, Collingwood takes his argument to completion where the fully engaged designer's work allows the total imaginative experience to be available to the spectator. Such practices eventually allow a cultivated intuition to see air.

Exercising the Material Imagination

Designers engaged in the productive arts constantly struggle to find the correct balance between accommodating practical requirements and inspiring human sentiment in their work. Neither alone is sufficient. For instance, architects distinguish between architecture and utilitarian buildings. While both provide shelter, have structural integrity and respond to pragmatic programs, buildings considered architecture have achieved a measure of emotional satisfaction perceptible in their experience. Becoming licensed, as is required for architects and landscape architects, follows years of education that nurture talent and creativity as well as a lengthy apprenticeship to develop technical mastery. One motive of design education is for the student to develop techniques of inquiry that support their nascent ideas by developing tools to help their work mature; that is, to acquire methods of formulating and materializing design intent. Social anthropologist Tim Ingold notes: "Western thought sees intentionality as residing not in the action itself but in a thought or plan that the mind places before the action" (Ingold 2000: 103). Thus located, the creative urge in the productive arts finds little appreciation when a designer's self-absorbed wilfulness produces projects requiring others to experience places in limiting and too prescriptive ways. On the other hand, ideas meant to serve others, once envisioned and cultivated, lead to open designs where a range of thoughts, impressions, activities and responses are possible. Such designs are successful because they are satisfying in the way they intimately involve the user and satiate to some degree their physical and emotional needs. Such engagement is part of engendering responsible stewardship, critical to lively and sustainable public places and projects.

Homophonetic words have the same pronunciation and different definitions. Coincidentally, in the English language, three words – sighting, siting and citing – have an unusual alliance in their role in design development. Individually and in deliberate sequence, these terms progressively identify steps useful for exercising the material imagination whose robust condition is critical to effective creative work. The intent of the exercise is to assist the imagination in thinking about physical materials independent from specific form. *What* is being considered is less important compared to *how* it is drawn to reveal its material qualities. Put another way, can the material imagination be productively active prior to the formal imagination? I suggest thinking *through* and *in* materials is a way of visualizing design materially. As this chapter hopes to demonstrate, when this is possible, design intent has the potential to transfer through physical materials to the experience of the completed project.

Sighting

There exists an extensive literature about visual theory applicable to the study of the productive arts (Crary 1992; Mitchell 2007, 2008; Stafford 2007; Elkins 2008). Seeing as a common, sensory and corporal operation as well as a cultural construct has also been well studied (Berger 1972, 1980, 1985; Foster 1999; Harris and Ruggles 2007). Sighting is a more intentional term for seeing and implies acts with directed awareness. Looking critically requires focus, discernment and evaluation. What are the qualities that make the object in view what it is and not something else? Perception involves all the senses yet the domination of the visual sense is unchallenged to the point that Ingold observes many "lay the ills of modern Western civilisation at the door of its alleged obsession with vision" (Ingold 2000: 246). Seeing is more than an optical operation; understanding what is seen is a thoughtful activity. Moving between the physical, experiential realm of the senses, especially vision, and meta-physical reflective realm of thought feeds the imagination both as a means to understand experience and, for designers, as a means to create places of enriched experience.

Sighting as the first step in exercising the material imagination requires deliberate looking which turns the activity from spectatorship to reading (Mitchell 1994: 16). With the sustained gaze, the designer's visual imagination explores possibilities now laden with vague impressions as well as certain conclusions. Training occurs by drawing what is seen without necessarily knowing the drawing subject or without the subject being important to the drawing's content, as illustrated in Figure 6.1, "From Imitation to Expression." Looking discriminates what is seen in a slow exchange between drawing lines and the thoughtful reflection about what is presented, and then re-presented. As Ingold observes quoting psychologist James Gibson, "learning is an 'education of attention'" (Ingold 2000: 416). Beyond the kinds of drawings made to record what is seen, these attentive drawings are made to explore what is possible. Everything material before your eyes contributes to this operation: the size, color and texture of the paper, the manipulation of the drawing instrument, as well as the way lines appear with their own material qualities of dimension, weight and edge. All procedural aspects alter the representational tone reflecting the material quality of the object being drawn. Drawing with materials in mind allows materiality to be more readily observable, a procedure amplified with practiced looking. Just as pitch is improved with vocal training, visual acuity improves with drawing, but these drawings need to be more than faithful imitations of the observable. Cartoonist Saul Steinberg worried that drawing from life told truths sometimes revealing an uncomfortable complicity for the artist in the questions raised (Steinberg 2002: 69–71). Intense sighting questions the deliberate correspondence between sensory impressions and drawings exploring possible relationships between idea and effect. Free from judging significance or value, such drawings investigate materiality because it cannot be avoided. Shapes take on physical properties by the material of their making. The reading mediates from one context to another; that is, between what

Figure 6.1   From Imitation to Expression
Source: Cristina Lewandowski, 2010

is seen and what is imagined. Maurice Merleau-Ponty claimed the subject in view becomes objectified through sight. "It is the mountain itself which from out there makes itself seen by the painter; it is the mountain that he interrogates with his gaze" (Merleau-Ponty 1964a: 166). Attentive sighting is further refined when its setting or siting becomes part of the imaginative action.

Siting

Siting, the second step, is the act of situating the object in view in a specific place or site. The boundaries of a place may be invisible such as with legal property lines, ambiguous where use blurs what is within or borrowed from beyond, or precise with easily read physical clues. Siting conditions affect the object and therefore its perception (Burns and Kahn 2005). Investigating the site or, put another way, the circumstances that resulted in the particularity of object placement, re-positions acts of sighting in ways beyond strictly focusing on physical appearance. Aesthetics as a philosophical subject has too often been incorrectly narrowed to the visual sense especially in judgment of the beautiful. More accurately, aesthetics is the full field of perception including not only the material and formal character of an object as seen, but also its performative aspects. Attending to the other senses – taste, smell and sound – depends on the object, site and proximity to the person. The haptic sense however, requires touching. Felt sensation through contact or tactile sensations, even sometimes thought of as a condition of territorialization, expands perception as spatial conditions become apparent. Those who argue for "the necessity of deterritorializing culture" (Pinney 2005: 268)[1] have surrendered the power of physical attachment people have to place through located objects (Chow 2004). An object's site is local, immediate and cumulative of direct and indirect causes allowing it to be materially present in its physical surroundings. The relevance of site is clarified if we consider for instance, how everyday objects displayed in museum settings take on new meaning, or how environmental or land artists such as Richard Long and Andy Goldsworthy find resonance using local materials disturbed to focus attention on the place made more physically present through their work.

Evaluating the situated presence of the object of our gaze requires apprehending information not necessarily present. An example is observing the rhythm of a repetitious pattern suggesting a likely continuation not yet evident. As anthropologist Daniel Miller in his book *Material Cultures: Why Some Things Matter*, introduces the fundamental question of the materiality of specific domains where things are more accurately perceived and more richly comprehended in context where the implications of character are appreciated (Miller 1998).

Observing the situated condition uncovers phenomenological relationships contributing much information to the aesthetic evaluation. Unlike most works

---

1 Of course, Gilles Deleuze was referring to language (Smith 1997: xlviii), and Arjun Appadurai was referring to things as traded commodities of value (Appadurai 1986).

of fine art, the productive building arts are constructed in specific places. Once built, many things beyond the project itself affect the experience of the design. For instance, in the designed landscape, feelings of orientation and opportunities for movement are conditions altering perception of place (Conan 2003). Invisible forces such as gravity, vital actions of growing plants most intensely apparent during spring, and what nineteenth-century landscape architect and theorist J. C. Loudon called the "fleeting" also affect experience (Loudon 1806: 363, 414, 422). The occasional presence of Loudon's fleeting, such as the directed prevailing breeze or attracted songbird, was a measure of sophisticated and pleasing landscape design because its control depended in part on the accidental consequences of siting. Social anthropologist Alfred Gell might suggest the resulting perceived sensations are enchanting because they, like all art-systems, "in the final analysis ... serve necessities which cannot be comprehended at the level of the individual human being, but only at the level of collectivities and their dynamics" (Gell 1992: 43). Sighting as sited gives objects material presence beyond their own medium. The material imagination benefits from deepened observational skills as the full physical and emotional impact of experience is enhanced. Nevertheless, to engage the intellect such observations need to be tempered with external referents.

Citing

The third step of the exercise takes gathered sensory impressions and evaluates possible implications. This activity assumes the object of our gaze is firmly linked to the site in its full perceptible materiality. Now their union can be considered in a more transcendent way. For example, considering a physical material in another spatial dimension, such as how it is depicted in paintings or described verbally, informs a deeper engagement because these disciplines are not constrained by the same limits as three-dimensional design that people physically experience. The material imagination evolves with reflective thinking about new relationships and associations.

Summoning external referents helps the designer because it reveals particular attributes that share universal qualities. Cultural practices accrue shared meaning because their significance has consistently emerged over time. For example, examining the circumstances at the time of an object's making and reviewing multiple reactions leads to imagining future possibilities. Summoning referents for design projects, projected or completed, has several possible directions such as inquiring about the designer's motive, the governing community's concerns or the constituent audience's responses. Making this information internal to the design investigation enhances understanding current conditions more fully.

This is particularly true with monument and memorial design, which often have conflicting programmatic agendas (Savage 2009). Narrative devices tell a particular story in a designed landscape urging a person to feel certain sensations empathically. Highly-charged design may actually transform the person to that emotional state. Labyrinths, for instance, were devoted to altering psychological states (Ball 1993: 37). Even an observer's memories change the impact of experience as meaning deepens when historic and cultural associations are recalled (Francis and Hester 1990). Showing traces of such associations in a drawing adds layers of meaning to the design investigation because over-laps emphasize possible connections over time and space. Pulling referents to the moment and making "it" mine is the consuming phase of ownership where the neutrality of possible associations ceases (Gell 1986: 113). This applies to non-physical referents too because as Miller observes "the large compass of materiality, the ephemeral, the imaginary, the biological, and the theoretical; all of that which would have been external to the simple definition of an artefact" comes into play (Miller 2005: 4). Citing requires finding and filtering information, and prioritizing choices which are activities reinforcing the material imagination.

The drawing in Figure 6.2, "An Imaginative Material Projection," shows earth surrounding the building "roots" heavily shaded to suggest stability. Showing contrasting materials in the drawing prompts speculation about earthy grounding and human aspiration with sharp, straight lines for man-made construction, and wispy lines that twist and then taper to nothing representing organic growth. The drawing wonders if vertical buildings grow out of the horizontal ground like trees. Referring to the material technologies of other disciplines, in this case with building footings shown as tree roots, strengthens design inquiry because the juxtaposition challenges conventional assumptions. W. J. T. Mitchell might say one stands for and acts as symptoms of what they signify (Mitchell 2005: 15). Natural materials

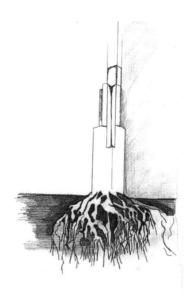

Figure 6.2   An Imaginative Material Projection
Source: Allison Scott, 2008

such as earth, wood and sky are precursors to the formed solid walls resulting in an object that is capable of participating in culture (Ingold 2000: 340–341). Materials as substances have no inherent meaning – they simply are – but as is concluded in this chapter, materials accrue significance when they are worked in ways where their materiality guides decisions about form. Sighted, sited and cited, the material imagination has the ability to engage and to cultivate artistic intuition. Thus, exercising the material imagination encourages the designer to set aside problem-solving approaches in favor of more open-ended inquiry. The question before us now asks if there are specific drawing techniques that develop ideas now made available with the primed material imagination. As previously mentioned, designers make representational drawings for the efficient conveyance of practical information. Designers also need to draw to explore embryonic ideas. These are generative drawings; ones meant to nurture ideas. Landscape architect and theorist James Corner distinguishes the demonstrative function of drawings as a "vehicle of realization" from the speculative function as a "vehicle of creativity" (Corner 2002: 159).

An interpretation of Collingwood's treatise on aesthetics in his *The Principles of Art* (1938) suggests four types of drawing aid such a speculative investigation. After considering Collingwood's argument for distinguishing between craft and art, the drawing techniques of material representation are examined starting with the literal and selectively literal, followed by the emotionally expressive, and concluding with drawing as a total activity of the complete imaginative experience.

Techniques of Material Representation

Collingwood defines craft (*techne*) as making an object where the result matches the desired intent, but does not exceed it (Collingwood 1938: 15–41). For architects who do not make buildings and landscape architects who do not make landscapes, their skilled craft is employed in representational demonstrative drawings needed for others to execute their design. Their competence is evident in well-crafted technical drawings. For designers, there is also much pleasure in making a rendered drawing with fine draughtsmanship, and it is also important to have some familiarity with the effort required to make a finished product of high quality, for their pleasure and satisfaction, and to better appreciate acts of crafting constructions. Mitchell observes there is little confusion discerning the nonartistic and artistic object although why a distinction is needed in anthropological studies is less clear (Mitchell 2005: 347–348). Technical drawings are meant to realize designs and their aim is not to be artistic, but rather efficient and clear. Design development drawings on the other hand, are artistic because their aim attempts to exceed expectations. Material culture anthropologist Fred Myers explores the difference between paintings made by indigenous cultures following custom and tradition, and the uniquely original work of fine art, and concludes each has a different intentional procedure (Myers 2005: 97–99).

All production, Collingwood notes, involves three things: materials, the means or tools to work the material, and the product made from the assembled parts (Collingwood 1938: 16, 17). If this sequence repeats, then each new product may be considered the "raw" material for another series of refinements. At first, materials are pure elements, substances found in nature such as carbon, oxygen and hydrogen. These elements combine as chemical compounds, and along with small amounts of ash for instance, make the wood of trees. Practices of cutting, dressing and joining works the material adding value as the wood is formed into a useful object, such as a chair (Ingold 2000: 80). The craftsman's process is successful when it leads to a pre-determined final result. Anthropologist Sir Raymond Firth criticizes Collingwood's early position in *Outlines of a Philosophy of Art* (1925) which sets natural beauty opposite to artificial beauty except "as a superficial distinction between degrees of involvement with the relationship in the material" (Firth 1994: 18). Collingwood clarified his definition of art in his later book (Collingwood 1938: 108, 132–135) where the difference became a matter of preconception.

Craft and art come together, for instance, in the furniture of master woodworker George Nakashima. His tables and chairs reflect an appreciation for every nuance of grain, burl, check and knot in the wood making well-crafted ordinary objects of unique beauty. It is as if the final form emerges from the character of the material itself. In the end, the complexity of the productive arts must blend craft in its executive and art in its inquiry.

The following four drawing techniques of material representation are arranged in order with increasing connection to human feeling. Ultimately, the dialogue between drawing and designer is vital to the questioning Collingwood considered fundamental to making art. These techniques aid imaginative thinking when applied to the productive arts expected to operate with purpose, and at the same time are appreciated for providing what art can provide.

Literal – Name | Rubbing | Photograph

The first type is a literal drawing with a one-to-one correspondence between physical material and drawn image. Its literality is manifested in a process of selecting a specific material that necessarily has a name, by drawing through rubbing transferring tactile qualities from material surface to paper, and by comparing the result to a photograph.

Words obtain their currency from accepted usage (Nuessel 1992). Linguistic conventions are handy when identifying things, and precise communication depends on accurate terminology. Word definitions evolve as society demands change reflecting practice, and these shifts increase the accuracy and ability to communicate reliably with one another. Moreover, examining shared etymological roots of terms reveal common properties often overlooked in everyday language. For instance, the words "theory" and "theatre" share the same base θεα – to look on, view, contemplate (*OED*). Knowing this reminds those concerned with theoretical issues that this work depends more on contemplative observation than active participation. As Miller explains, a staged performance provides the frame or context along with clues that govern actor and spectator behaviour, especially urging audience restraint to not interfere and to let the story unfold (Miller 2005: 5). So too theoretical investigations become engaged by examination art historian Michael Fried likens to "literalist work [that] depends on the beholder, is incomplete without him, it *has* been waiting for him" (Fried 1998: 163). A precise name is a verbal convention; a precise image can become a visual convention.

Rubbing the surface of an object with a soft pencil over thin paper allows the distinct texture of the material to emerge as something to see.[2] In the left panel of Figure 6.3, the pine tree bark rubbing reveals its particular growth habit, pattern and proportional scale of face and fissure, and nothing else. Like a minimalist painting or sculpture, Fried says "literalist art defines or locates the position it aspires to occupy" (Fried 1998: 149). The drawing records descriptive qualities that are the material's modifying adjectives. It also suggests something of the bark's character which comes through as if it were modifying a way of being, here more rough than smooth, more vertical than horizontal, and hard yet able to deflect slightly. In this way, the drawing is generally "woody" and then specifically "piney" or pine-like. In the rubbed drawing, there is no attempt to create an effect, to assign meaning, or to alter the appearance in any way thus making it a literal representation. Collingwood would consider the result well-crafted because the

Pine - Pinus

Figure 6.3   Literal and selectively literal
Source: Keri Kennedy, 2009

---

2   Many memorials include the names of those being remembered. For instance, at the Vietnam Veterans War Memorial in Washington, D.C., family and friends find comfort in making a rubbing of a name inscribed on the permanent stone wall surface.

person making the drawing has the single intent of participating in the feel and then the look of the material surface (Collingwood 1938: 45).[3] Gell might admit that similar to ritual type drawings where "graphic gestures do not represent something, but constitute something," the content of a rubbing is the relationship between the object and the person making the drawing (Gell 1998: 191).

Rubbings take time and it is tempting to snap a quick photograph instead. Roland Barthes declared "only the photograph is able to transmit the (literal) information without forming it by means of discontinuous signs and rules of transformation" (Barthes 1984: 43). In the middle panel of Figure 6.3, a photograph of the tree bark is printed at full scale which helps the image appear only as unformed material. The photograph provides an exact duplication of the surface appearance. Art critic John Berger claims "no painting or drawing, however naturalistic, *belongs* to its subject in the way that a photograph does" (Berger 1980: 50). The photograph seems to yield information equivalent to the rubbed drawing, but it must be different because it comes into visual practice through a different social process. The hand making the tree bark rubbing holds paper and instrument, and rubs back and forth until the drawing is complete, about which more later. The artist-less photographic record (Gell 1998: 35) finds its agency in its availability for recontexualization (Rose 2007: 216–225).

There is no doubt a photograph is useful for identification and documentation where the physical "moment" is captured with objective validity, but it does not allow the viewer to develop much feeling toward what is seen because the record reflects little subjective contribution besides the decision to take the photograph. Although under other circumstances, the photographer displays something unnoticed, here the conditions are already present and the photographer does not pose them or look to capture a "decisive moment" to use Henri Cartier-Bresson's term. Rather, looking past appearance occurs with the feel of the pencil moving over the paper allowing material qualities to emerge. As Berger notes, "the photograph stops time, while the "drawing or painting forces us to *stop* and enter its time" (Berger 1985: 149).

Selectively Literal – Essential Characteristics Only

For the second type of representational drawing, shown in Figure 6.3 as the right panel, the tree bark rubbing is redrawn with lines and shading. Mitchell compares this type of effort to drawing water from a well; that is, each movement is considered prior to execution (Mitchell 2005: 59). Decisions are made to avoid repetition and to show, to use Collingwood's words, the "bold selection of important or characteristic features and suppression of all else" (Collingwood

---

3 Collingwood's philosophical activities had an archaeological side. He spent time in the field making tracings of Roman inscriptions found in Britain's ancient ruins. He contributed to a book with J. N. L. Myers called *Roman Britain and the English Settlement (1936)*, part of the Oxford History of England Series.

1938: 54, 55). Minimal types of drawing suggest their source by having only sufficient lines as needed to trigger identification (Linsky 1967). Their purpose is to isolate and thereby increase awareness of relevant attributes. Even at its most abstract as Mitchell proposes, "liberated from image-making" these kinds of drawings prime the imagination with familiar clues (Mitchell 2005: 44). Selectively literal drawings do not symbolize actual physical materials; rather they represent materials in the most open possible way. The drawing is complete when recognition is possible because the seer's imagination acts to fill in the blanks. The abstracted drawing privileges the essential over the ornamental, and requires the mind to complete the prompt (Walker 1997). In its simpler state, selectively literal drawings reach deeper into the unconscious bringing forth reflective response to the material as shown. Even though the material's surface is taken from life, recognition depends both on the artist's memory and the seer's experience which leads them to surmise that this drawing is of tree bark. A selectively literal drawing presents an opportunity to visualize materials more in accordance with their true essence than when literally represented because the seer is more engaged in understanding the seeing. Its presentation acquires presence in concert with the beholder's increasing awareness.

Collingwood thought literal and selectively literal drawings could successfully convey emotional effect although this was not their aim (Collingwood 1938: 54). Thus they were more a product of technically proficient craft than art. Yes, such drawings are clear and able to designate specific information unambiguously, but they are limited by the specific reason for which they were made. As material records, such unformed images that contain no human determination cannot be expected to contribute to a community's material culture. They were never meant to express, comment on or contain its cultural identity (Miller 1998: 16, 17). The issue here is that literal pictorial description is not expression because it does not attempt emotional content (Collingwood 1938: 238).

Emotional Representation – Drawing from and to a State of Mind

Another type of representational drawing aims at evoking emotional response bypassing description and appealing directly to the expression of feeling. These are no longer drawings attempting to record or resemble an appearance, but rather drawings with the specific intent to represent emotion through materials. Drawing from and to a state of mind begins with an artist being aware of their primary emotional state, suppressing other feelings, and using that energy to explore the emotive quality in a particular material.

Prior to making the drawing shown as Figure 6.4, the artist identified different types of stone, made rubbings of their unfinished surfaces, and photographed their appearance. Using a field stone footbridge as subject, the artist made a drawing after first isolating a particular emotion felt at that time, in this particular case the feeling was "uneasy." Such expression individualizes the drawing subject because

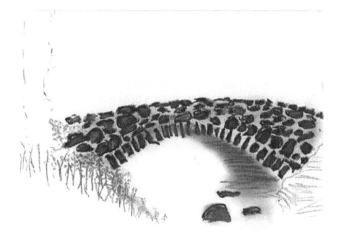

Figure 6.4   Emotional representation: an uneasy feeling
Source: Michael Luk, 2009

it shows a way an object may be seen through a determined and distinct vocabulary of expression. Berger says through drawing,

> each confirmation or denial brings you closer to the object, until finally you are, as it were, inside it: the contours you have drawn no longer marking the edge of what you have seen, but the edge of what you have become (Berger 2005: 3).

As shown, the stones in the arch appear to hold each other up, and the nearly absent mortar conveys a sense of uncanny suspension. The stones, drawn solid, appear unsupported and a feeling of uneasiness transfers from artist to viewer. The place where the bridge meets land is also vaguely implied, increasing the unsettled feeling. Drawings expressing an artist's emotions depend on two things. First, they come from their state of mind where expression is not a cathartic expulsion of irrational emotions, but more of a way of drawing with directed intent. They also depend on thinking about a particular thing more through its materiality than through its form. The question is not if the object in the drawing is a footbridge. Drawings that represent emotion take advantage of a material's physical and associative properties in order to make a picture that conveys both that understanding and the artist's particular sentiment. Ingold links feeling and being "in touch" with the world (Ingold 2000: 23).

What is the source of an artist's understanding supporting this effort? Here, it requires knowing rock used for stone construction is cut into recognizable shapes and sizes with dressed surface textures. Further, stone is hard and has weight giving it structural capability. Knowing something of stonework construction techniques adds to the possible visual associations because stone construction is held in place with removable scaffolding until the wet mortar dries achieving the required compressive strength and lasting solidity. Stonework which is placed individually never quite relinquishes an echo of its piece-by-piece assembly. Stone constructions convey a sense of permanence because we know of stonework not only much older than ourselves, but also likely to outlast us. All this information combines into an emotional reaction stimulated when seen in such a drawing.

If the intent is to express an understanding of the subject satisfying an internal question only, then the viewers' corresponding response is irrelevant. Rather, drawings of emotional representation are meant to help the artist embed their work with particular sentiments likely to be felt by others. This then, leads our inquiry to question if drawing materially conveying specific emotions transfers to designing places which evoke that the same feeling.

Considering early American public park design is a way to begin addressing the relationship between material choice and design (Karasov and Waryan 1993; Tate 2001; Czerniak and Hargreaves 2007). Architectural sociologist Galen Cranz traces the history as park types changed from pleasure grounds meant to relieve over-crowded, dirty, noisy and disease-prone cities to reform parks of ideal bourgeois leisure to recreational facilities, and notes the type and placement of plants, furniture and equipment changed too (Cranz 1982). The design intent of specific places within large parks was reinforced with the choice and execution of materials. In Central Park for instance, intimate settings "small enough to be taken in at a single glance" (Miller 2003: 137) are meant to promote feelings of quiet reflection. Structures were made of wood used in its rustic condition left untrimmed and retaining its bark surface joined with angular shapes. In the large places with grand visual sweep, noble sentiments were displayed with refined stone materials finished with smooth, polished surfaces accompanied by sculptural iconography in fine detail. To link the scale of place, intimate or grand, with materials either rustic or refined, was to encourage sentiments subsumed by nature or celebrating cultural achievement. Drawings that achieve designs such as this have devices that can be employed to achieve emotive response. For instance, for an image to evoke a desired base emotion such as dread, its subject cannot just be called dreadful. The feeling of actual dread has to be aroused in the artist and the audience. Suggestive visible attributes such as the exaggerated and intimidating size of a wild animal's fangs is one way to evoke a feeling of dread in painting. Another approach is to use a tone of apprehension and tension perceptible in material representation which will just as surely bring forth certain sentiments. Merleau-Ponty said base emotions do not look to mental references; that emotion itself is expressed in response to the apparent threat (Merleau-Ponty 1962: 184).

These kinds of drawings, however, only begin to consider techniques for how designers can draw expressively. Relying exclusively on emotional expression risks design results others find uninteresting; they simply have little desire to share that emotion or they feel unconnected to places that dictate a narrow emotional response. Promoting a completely sympathetic dialogue between projective representational drawings and intended response requires more.

Total Activity – Seen as Felt

Collingwood illustrated his thoughts about why painting in the twentieth-century is more than a visual art with the work of Cézanne. Mitchell finds general acceptance that "vision is a 'cultural construction,' that is learned and cultivated, not simply given by nature ... and that it is deeply involved with human societies, with the ethics and politics, aesthetics and epistemology of seeing and being seen" (Mitchell 2005: 337–338). Beyond art as visual, art historian T. J. Clark argues "Cézanne's dogged attention to sensory fact" (Clark 2001: 94) was an attempt to capture a single moment in a single view in all its complexity. D. H. Lawrence explains the advantage succinctly:

> I am convinced that what Cézanne himself wanted *was* representation. He *wanted* true-to-life representation. Only he wanted it *more* true to life. And once you have got photography, it is a very, very difficult thing to get representation *more* true-to-life: which it has to be (Lawrence 1929: 254).

Collingwood admired the way Cézanne painted more than the appearance of what he saw and indeed, more than the feeling he wanted to express. Cézanne painted in ways where the work is actively open to emotion. Clearly, Cézanne's paintings were composed to capture the felt experience and not to record the visually-faithful appearance. Collingwood says of this work "his landscapes have lost almost every trace of visuality. Trees never looked like that; that is how they feel to a man who encounters them with his eyes shut" (Collingwood 1938: 144). The scene is not just pictured. It is also expressed materially through the way Cézanne builds up color and brush stroke, painting what he feels for the subject. This technique paints the way the visual may be felt. Ingold observes "visual space is presented to the sighted all at once, but tactile space has to be assembled by the blind, bit by bit, through a repetitive and time-consuming exploration with the fingers" (Ingold 2000: 271). An account of how this happens in Cézanne's work suggests a technique for practicing Collingwood's final aesthetic category of total activity.

Collingwood concluded his chapter on art and the imagination saying "thus a work of art proper is a total activity which the person enjoying it apprehends, or is conscious of, by the use of his imagination" (Collingwood 1938: 151). This is the fullest possible engagement – a total imaginative experience – between artist and audience where materiality is shown in ways urging full affective response. For this to happen, the seer's total body must be engaged, the limbs in particular.

Over thirty years, Cézanne painted the same mountain producing at least forty-four oils on canvas and forty-three watercolors on paper. Collingwood says Cézanne "broods the obsession of Mont Sainte-Victoire, never looked at, but always felt, as a child feels the table over the back of its head" (Collingwood 1938: 144). Clark describes Cézanne's mountain painted "bit by bit" as, the blues are translucent, floating into and over the answering parallelograms of green. The mountain looks crystalline, made of a substance not quite opaque, not quite diaphanous; natural, obviously, but having many of the characteristics – the crumbled look, the piecemeal unevenness – of an object put together by hand (Clark 2001: 95).

Here, looking at this Cézanne painting in Figure 6.5, memories of climbing mountains enter your imagination. You remember feeling the ache of tired feet unaccustomed to uneven rocky paths. Your legs feel the fatigue from the effort to ascend. A bit winded, heart pounding, anticipation builds expecting scenic reward for the muscular effort. These sensations are tapped in Cézanne's selected vantage point overlooking a valley and viewing the distant mountain. He positions the spectator spatially separated from the object of his gaze with the tree in the foreground constructing the space between you and the mountain, although,

as later photographic researchers at the spots at which Cézanne set up his easel were to prove – showed more of Mont Sainte-Victoire than could ever actually be seen from any *one* of these vantage-points. These landscapes depict, not any fixed appearance of Mont Sainte-Victoire, but Cézanne's interaction with this object over time ... (it) is revealed as a *process*, a movement of *durée*, rather than as a 'thing' (Gell 1998: 244).

Figure 6.5   Paul Cézanne, Montagne Sainte-Victoire (66.8 × 92.3 cm), oil on canvas, c. 1882

Source: The Samuel Courtauld Trust, The Courtauld Gallery, London

The tree is also painted in a way meant to engage your limbs. Your hands sense the rough, woody surface and you feel as if your arms could wrap around the tree trunk. Cézanne engenders this feeling by painting the tree with strong outlines setting it as an object spatially apart from the background making what he called a silhouette (Simms 2008: 77). Thus distinguished, an emotional response is elicited to draw near, to touch and to allow the felt material to enter our imagination through our hands. Berger, in a passage comparing photography to drawings that encompass time, says we are not just seeing a tree as an object, but as "a tree-being-looked-at" where the material representation accrues definition fusing appearance with remembered experience (Berger 1985: 146–151).

Cézanne's intent also reached beyond the visual and haptic senses. He wanted the viewer to engage all their senses in a full aesthetic commitment. Cézanne said about his work,

> the smell of the pines, which is harsh in the sun must be matched with the green smell of the grass, the smell of the stones, and the smell of the distant marble of Mont Sainte-Victoire. It is this that must be rendered: and only using colors, without literature (Lapenta 2005: 168).

Capturing the total imaginative experience of a physical place in a flat surface is difficult, and it takes a Cézanne to achieve it. For the productive arts, the target is reversed. Cézanne painted actual landscapes in two-dimensions; designers draw in two-dimensions (even three-dimensional computer simulations are "drawn" on a flat screen) in order to imagine and to design three-dimensional projects and places. Nevertheless, Cézanne's painting illustrates how a dialogue occurs between what is seen and what is felt.

Another point Collingwood explores is how artists knows their work is complete. Rilke noted the presence of blank or empty space when he examined Cézanne's paintings for a sustained time during a memorial exhibition in Paris in 1907. He suggests Cézanne painted what he knew, and left blank what he didn't (Rilke 1952: 46). The white watercolor paper left untouched, the reserve, is evidence of such thinking, or maybe it reflects Cézanne's approach to holding open space for the imagination. Mitchell claims what the picture lacks is what it wants (Mitchell 2005: 25). Baudelaire on Cézanne observed "There is a great difference between a work that is complete (*complète*) and a work that is finished (*fait*)" (Simms 2008: 92). Cézanne appreciated the distinction and said a finished painting "gives rise to the admiration of idiots," while "I must instead only seek completeness for the pleasure of working in a more truthful and sound manner" (Simms 2008: 94). Knowledgeable, experienced and with a heightened sense of awareness, the artist can "tell" when their work is complete noticing both the state of the object and the circumstances of its making. Another word for this kind of sensitivity and responsiveness is *intuition* (Ingold 2000: 24, 25, 190).

For the productive arts, working in a more truthful and sound manner requires a dialogue between talking drawings and listening designers. The work goes through many rounds of proposing questions and responding. Work on design idea drawings continues until a measure of satisfaction is achieved, communication quiets and the investigation is complete. The drawings in Figure 6.6 show a visual and verbal vocabulary used later for a park design. The landscape elements include meadow, sea coast and woods with people in various engagements. The ascribed feelings are linked to places phenomenologically where one is meant to feel exposed, isolated or enclosed, where a prospect is revealed, or where one feels a sense of solitude. Emotive proposals are coupled with building materials such as earth that is mounded up, levelled or cut into. Concrete walls retain, trees frame or provide a sheltering canopy, and paved paths encourage human movement through the woods. All of these elements propose choreographed relationships between the materiality of place and experience. As Clark concludes "no wonder we can never be sure where materiality ends and phenomenality begins. Each thrives interminably on the other's images and procedures" (Clark 2001: 99). These vignettes of proposed experience then become the "raw" material for later design decisions about form and composition. Collingwood's total imaginative experience now is possible following a dialogue between material and project. In this way, expression has given way to indication.

The speculative drawings in Figure 6.6 can then guide demonstrative drawings prepared for construction. Even during the building phase, the designer still has the opportunity to reflect on the progress, refer back to original design intentions, and fine-tune as needed. And in its next cycle of existence, the finished work is available for others to experience who determine its success based on its emotional appeal as well as its practical competence. The project then begins a new phase of dialogues because the physical building materials continue to change and to age prompting new perceptions and suggesting new associations extending the total imaginative experience indefinitely. A physical dialogue between material and exposure becomes visible through weathering. In processes of addition and subtraction, metals acquire patina, wood changes color and texture as light is absorbed, and moss and other plants take root and grow in tiny pockets of masonry where soil and moisture collect. By subtraction "finished corners, surfaces, and colors are 'taken away' by rain, wind, and sun" (Mostafavi and Leatherbarrow 1993: 6). Use patterns also emerge over time as hands repeatedly move along brass or bronze railings polishing part of the surface, feet continue to wear away sharply-cut granite step edges, and the center of paths are compacted into shallow swales. It is as if the traces of past experiences, Foucault's histories, gradually emerge reinforcing a material's continued life and inviting further participation in a similar manner (Foucault 1994: 370–371). A person's experience in that moment has an opportunity to participate in a place's material history. Ingold says "places do not have locations but histories" (Ingold 2000: 219). To participate in the material culture of that time in that place is to be touched by it, where your hand and your foot follows where many hands and many feet have gone before, and where others will continue to come.

...*extending over the meadows...*
exposure

...*and the extensive prospect of the sea coast...*
reveal

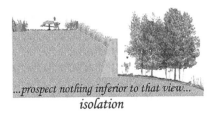

...*prospect nothing inferior to that view...*
isolation

... *views may be seen either distinctlty or blended together...*
enclosure

...*from whence the prospect is terminated by the woods...*
solitude

Figure 6.6   Total activity
Source: Keri Kennedy, 2009

Conclusions

Different drawing techniques convey feelings through material visualization to different degrees. Drawings that represent emotions find added meaning when such expression extends to a material's associated haptic properties such as weight, thinness, smoothness or delicacy. When what is seen is felt, drawings bring to consciousness more of the design intent. Working through the material imagination allows the designer to set aside temporarily pragmatic judgments, programs and evaluations for the exploration of the material itself, and the various ways it can be worked. Designers are then better prepared to transfer those attributes to the intended lived experience of the design project. This action continues investigating aspects of design that address purpose in the way materials are shaped into form, and then assembled into a composition. In other words, the design practice becomes in-formed through material comprehension. Collingwood noted the difference between material and form saying "the matter is what is identical in the raw material and the finished product; the form is what is different, what the exercise of the craft changes" (Collingwood 1938: 16). The matter of material persists as materiality and honing the visual imagination through speculative drawings taps the necessary thoughtful engagement between intention and experience needed for good design work.

Again, there is no doubt that part of the competent design professional's responsibility is to communicate information clearly for execution. Nevertheless, before this is done well, the *designer* as *artist* must explore ideas with heightened awareness and testing sentiments intended for a particular design to be used by others. Such drawings translate dreamy whispers into design statements. Collingwood's argument requires the artistic consciousness be uncorrupted as this is critical to the proposal's moral integrity and the corresponding clarity of the on-going dialogue (Jones 1972: 46; Dreisbach 2009). Collingwood's artist cannot be free from intention because only embodied materials allow objects to become available to meaningful perception. The kind of questions and the kind of dialogues between artist and object through drawing tempers the advancing intuition and its productive result.

Of all the materials – modern and ancient agents of change – air is the medium of free movement. Gaston Bachelard declared "where there is no movement there is no matter" (Bachelard 2002: 8). As such, the artist's material imagination is active only with the free movement between idea and drawing. Mitchell calls air "the necessarily invisible medium through which all spatial perception must travel" (Mitchell 2007: 44). When visual thinking evolves to the point that air is seen or visualized, then the material imagination perceives space as well as that which contains and defines place. Art historian Matthew Simms writes Cézanne had a term called the "envelope" for "the reflections and refractions of light that take place in the atmosphere that intervenes between the artist and his motif" (Simms 2008: 127). To see the invisible envelope is to be keenly aware of what mediates between visuality and materiality. For the designer, the implication of perception is internalized and becomes the basis for asking the key question of design: what is possible? Moreover, the designer must find a way to cultivate this question with a dialogue oscillating between drawing and idea. The designer must find a way to make a drawing talk and find a way to listen, to respond and to continue the reciprocal questioning in order to transfer intent to completed proposal. This then becomes the place for continued conversations between the built design and the experiencing user.

Ethnologist Howard Morphy notes "while an artist familiar with his or her audience may try to create an object which elicits a particular response from the audience, the creator of an object is never the complete master of its aesthetic potential" (Morphy 1992: 182). The work therefore must be both directed and open. Drawing through materials keeps the design inquiry open. Designers as artists explore through material attributes and their performative capabilities seeking the means to design for the total imaginative experience possible with uncorrupted intent. The results allow people to feel more consciously the secrets of their own hearts because they are experiencing projects meant to speak in ways that can be heard visually. Exercising the material imagination and developing drawing techniques of material representation engages and cultivates artistic intuition, making visible indications of human life in the ways we can be in the world.

Acknowledgements

I thank Jack Sullivan, FASLA for constructive comments on an early draft, Magenta Livengood and Colette Silver for close readings of later drafts, and the editors for their helpful suggestions. I particularly thank the students who confidently went on this investigation with me even though we weren't quite sure where it might lead.

References

Appadurai, A. (ed.) (1986), *The Social Life of Things: Commodities in Cultural Perspective*. (Cambridge: Cambridge University Press).

Bachelard, G. (2002), *Air and Dreams: An Essay on the Imagination of Movement*, translated by E.R. Farrell and C.F. Farrell. (Dallas: Dallas Institute).

Ball, E. (1993), To Theme or Not to Theme: Disneyfication without Guilt, in *The Once and Future Park*, edited by D. Karasov and S. Waryan. (New York: Princeton Architectural Press).

Barthes, R. (1984), *Image – Music – Text*, translated by S. Heath. (London: Fontana).

Berger, J. (1972), *Ways of Seeing*. (London: Penguin Books).

Berger, J. (1980), *About Looking*. (New York: Pantheon Books).

Berger, J. (1985), *The Sense of Sight*. (New York: Pantheon Books).

Berger, J. (2005), *Berger on Drawing*, edited by J. Savage. (Aghabullogue: Occasional Press).

Brown, B. (ed.) (2004), *Things*. (Chicago and London: University of Chicago Press).

Burns, C.J. and Kahn, A. (eds) (2005), *Site Matters: Design Concepts, Histories and Strategies*. (New York and London: Routledge).

Chow, R. (2004), Fateful Attachments: On Collecting, Fidelity and Lao She, in *Things*, edited by B. Brown. (Chicago and London: University of Chicago Press), 362–380.

Clark, T.J. (2001), Phenomenality and Materiality in Cézanne, in *Material Events: Paul de Man and the Afterlife of Theory*, edited by T. Cohen, B. Cohen, J.H. Miller and A. Warminski. (Minneapolis and London: University of Minnesota Press), 93–113.

Collingwood, R.G. (1938), *The Principles of Art*, 1977 reprint. (New York: Oxford University Press).

Conan, M. (2003), *Landscape Design and the Experience of Motion*. (Washington, DC: Dumbarton Oaks Research Library and Collection).

Corner, J. (1992), Representation and Landscape: Drawing and Making in the Landscape Medium, *Word & Image* 8: 3, 243–275. Reprinted in S. Swaffield (ed.) (2002), *Theory in Landscape Architecture: A Reader*. (Philadelphia: University of Pennsylvania Press), 144–165.

Cranz, G. (1982), *The Politics of Park Design: A History of Urban Parks in America*. (Cambridge, MA and London: MIT Press).

Crary, J. (1992), *Techniques of the Observer: On Vision and Modernity in the Nineteenth Century*. (Cambridge, MA and London: MIT Press).

Czerniak, J. and Hargreaves, G. (eds) (2007), *Large Parks*. (New York: Princeton Architectural Press).

Dreisbach, C. (2009), *Collingwood on the Moral Principles of Art*. (Selinsgrove: Susquehanna University Press).

Elkins, J. (ed.) (2008), *Visual Literacy*. (New York and London: Routledge).

Firth, R. (1994), Art and Anthropology, in *Anthropology, Art, and Aesthetics*, edited by J. Coote and A. Shelton. (Oxford: Clarendon Press), 15–39.

Foster, H. (ed.) (1999), *Vision and Visuality*. (New York: New Press).

Foucault, M. (1994), *The Order of Things: An Archaeology of Human Sciences*. (New York: Vintage Books).

Francis, M. and Hester, Jr., R.T. (eds) (1990), *The Meaning of Gardens: Idea, Place, and Action*. (Cambridge, MA and London: MIT Press).

Fried, M. (1998), *Art and Objecthood: Essays and Reviews*. (Chicago and London: University of Chicago Press).

Garner, S. (ed.) (2008), *Writing on Drawing: Essays on Drawing Practice and Research*. (Chicago: University of Chicago Press).

Gell, A. (1986), Newcomers to the World of Goods: Consumption among the Muria Gonds, in *The Social Life of Things: Commodities in Cultural Perspective*, edited by A. Appadurai. (Cambridge: Cambridge University Press), 110–138.

Gell, A. (1992), The Technology of Enchantment and the Enchantment of Technology, in *Anthropology, Art, and Aesthetics*, edited by J. Coote and A. Shelton. (Oxford: Clarendon Press), 40–63.

Gell, A. (1998), *Art and Agency: An Anthropological Theory*. (Oxford: Clarendon Press).

Harris, D. and Ruggles, D.F. (eds) (2007), *Sites Unseen: Landscape and Vision*. (Pittsburgh: University of Pittsburgh Press).

Ingold, T. (2000), *The Perception of the Environment: Essays in Livelihood, Dwelling and Skill*. (London and New York: Routledge).

Jones, P. (1972), A Critical Outline of Collingwood's Philosophy of Art, in *Critical Essays on the Philosophy of R.G. Collingwood*, edited by M. Krausz. (Oxford: Clarendon Press), 42–67.

Karasov, D. and Waryan, S. (eds) (1993), *The Once and Future Park*. (New York: Princeton Architectural Press).

Küchler, S. (2005), Materiality and Cognition: The Changing Face of Things, in *Materiality*, edited by D. Miller. (Durham and London: Duke University Press), 206–230.

Lapenta, S. (2005), Mont Sainte-Victoire, in *Cézanne*, preface by A. Gatto. (New York: Rizzoli), 168–169.

Lawrence, D.H. (1929), Introduction to these Paintings, in *The Paintings of D.H. Lawrence*, reprinted in *The Grove Book of Art Writing*, 1998 edition, edited by M. Gayford and K. Wright. (New York: Grove Press), 253–258.

Linsky, L. (1967), *Referring*. (New York and London: Routledge & Kegan Paul).

Loudon, J.C. (1806), *A Treatise on Forming, Improving and Managing Country Residences, and on the Choice of Situations Appropriate to every Class of Purchaser …* . (2 vol. London: Longman, Hurst, Rees, and Orme, Paternoster-Row).

Merleau-Ponty, M. (1962), *Phenomenology of Perception*, translated by C. Smith. (New York and London: Routledge & Kegan Paul).

Merleau-Ponty, M. (1964a), Eye and Mind, translated by C. Dallery, in *The Primacy of Perception, and Other Essays on Phenomenological Psychology, the Philosophy of Art, History and Politics,* edited by J.M. Edie. (Evanston: Northwestern University Press), 159–190.

Merleau-Ponty, M. (1964b), Cézanne's Doubt, in *Sense and Non-Sense*, translated by H.L. Dreyfus and P.A. Dreyfus. (Evanston: Northwestern University Press), 9–25.

Miller, D. (ed.) (1998), *Material Cultures: Why Some Things Matter*. (Chicago: University of Chicago Press).

Miller, D. (ed.) (2005), *Materiality*. (Durham and London: Duke University Press).

Miller, S.C. (2003), *Central Park: An American Masterpiece*. (New York: Harry N. Abrams in association with the Central Park Conservancy).

Mitchell, W.J.T. (1994), *Picture Theory*. (Chicago and London: University of Chicago Press).

Mitchell, W.J.T. (2005), *What do Pictures Want? The Lives and Loves of Images*. (Chicago and London: University of Chicago Press).

Mitchell, W.J.T. (2007), Landscape and Invisibility: Gilo's Wall and Christo's Gates, in *Sites Unseen: Landscape and Vision*, edited by D. Harris and D.F. Ruggles. (Pittsburgh: University of Pittsburgh Press), 33–44.

Mitchell, W.J.T. (2008), Visual Literacy or Literary Visualcy?, in *Visual Literacy*, edited by J. Elkins. (New York and London: Routledge), 11–30.

Morphy, H. (1992), From Dull to Brilliant: The Aesthetics of Spiritual Power among the Yolngu, in *Anthropology, Art, and Aesthetics*, edited by J. Coote and A. Shelton. (Oxford: Clarendon Press), 181–208.

Mostafavi, M. and Leatherbarrow, D. (1993), *On Weathering: The Life of Buildings in Time*. (Cambridge, MA and London: MIT Press).

Myers, F. (2005), Some Properties of Art and Culture: Ontologies of the Image and Economies of Exchange, in *Materiality*, edited by D. Miller. (Durham and London: Duke University Press), 88–117.

Nuessel, F. (1992), *The Study of Names: A Guide to the Principles and Topics*. (Westport and London: Greenwood Press).

Pinney, C. (2005), Things Happen: Or, From Which Moment Does That Object Come?, in *Materiality*, edited by D. Miller. (Durham and London: Duke University Press), 256–272.

Rilke, R.M. (1952), *Letters on Cézanne*, edited by C. Rilke, 1985 edition, translated by J. Agee. (New York: Fromm International Publishing).

Rose, G. (2007), 2nd ed. *Visual Methodologies: An Introduction to the Interpretation of Visual Materials*. (London: Sage).

Rowlands, M. (2005), A Materialist Approach to Materiality, in *Materiality*, edited by D. Miller. (Durham and London: Duke University Press), 72–87.

Savage, K. (2009), *Monumental Wars: Washington, D.C., the National Mall, and the Transformation of the Memorial Landscape*. (Berkeley: University of California Press).

Simms, M. (2008), *Cézanne's Watercolors: Between Drawing and Painting*. (New Haven and London: Yale University Press).

Smith, D.W. (1997), Introduction, in *G. Deleuze, Essays Critical and Clinical*, translated by D.W. Smith and M.A. Greco. (Minneapolis: University of Minnesota Press).

Stafford, B.M. (2007), *Echo Objects: The Cognitive Work of Images*. (Chicago and London: University of Chicago Press).

Steinberg, S. and Buzzi, A. (2002), *Reflections and Shadows*, translated by J. Shepley. (New York: Random House).

Tate, A. (2001), *Great City Park*. (London and New York: Spon Press).

Walker, P. (1997), Classicism, Modernism, and Minimalism, in *The Landscape, in Peter Walter Minimalist Gardens*, text by L. Levy. (Washington, DC and Cambridge, MA: Spacemaker Press).

Wollheim, R. (1972), On an Alleged Inconsistency in Collingwood's Aesthetic, in *Critical Essays on the Philosophy of R.G. Collingwood*, edited by M. Krausz. (Oxford: Clarendon Press), 68–78.

Chapter 7

Intra-actions in Loweswater, Cumbria: New Collectives, Blue-Green Algae, and the Visualisation of Invisible Presences Through Sound and Science

Judith Tsouvalis, Claire Waterton and Ian J. Winfield

... there *is* loss, as the mobile planet, human and nonhuman, continues on its way. There is *material* loss (things will disappear as they are reabsorbed into the cycles of destruction and creation); and there will also on occasions be a *sense of loss*. [...] ... [S]uch senses of loss cannot justifiably be mobilized as legitimations in themselves of a political strategy to reject change.

Massey, 2006 (emphasis in the original)

Well, I mean, if we have a dead lake we've got a dead community
Interview with Loweswater farmer, March, 2008

Introduction

This chapter considers the challenges posed by approaching environmental changes as an opportunity to do 'politics with things' (Latour, 2004). We describe how one such opportunity that arose in a specific setting, an upland English lake, Loweswater, which experienced increasingly frequent occurrences of (potentially toxic) blue-green algal blooms, was taken up supported by Science and Technology Studies (STS) theory and feminist philosophies of material intra-action, particularly the work of Barad (2003, 2007). To do politics with things things have to be materially present in the political process, something they often achieve – especially in environmental politics – in form of scientific visual representations (e.g. pie-charts, echograms, statistical tables, photographs, films, and other forms of media). These media are predominantly produced, studied, articulated, and mapped back onto the 'real world' by scientists in various settings, including political ones. Chemical compounds, biological organisms, hydrological- and geomorphological processes, and other 'things' such as those falling under the rubric of 'climate variables' become dots, lines, numbers and images on paper, slides, and in computer models. A dominant perception in the past has been to take such representations at face value and consider them as faithful mirrors of the real-world phenomena they represent. However, for many years now this simplistic linear epistemological and ontological understanding (from 'real-world' entity to faithful representation to straight-forward explanation and clear-cut action) has been critically scrutinized and undermined by STS scholars, critical visual theorists, and those advocating the necessity of a 'material' turn.

This chapter contributes to this critical field through a case study of the making of scientific representations to address a particular environmental problem or set of problems, and in line with the spirit of this edited volume, it attempts to show how representations come to be powerful actants in their own right – especially in political processes – and how their performative power shapes the becoming of the worlds they are embroiled in. As we go on to show, such representations co-produce knowledge, debates, decision-making processes, and perceptions of contested issues. Following Barad's (2007) performative metaphysics, visual scientific representations are understood here as momentary manifestations of a mutual entanglement of phenomena and the material–discursive practices that impose boundaries on them. They are a co-product of a mutual mingling of those trying to find out about the unknown, of the 'things' or phenomena that they come to know, of the apparatuses involved in the co-production of knowledge about such phenomena, and of the material–discursive practices mobilised to give such phenomena a definite form as 'this' or 'that' thing.

The chapter contributes to the aims of the present volume by describing and exploring a real-world experiment that put into practice a reconceptualisation of the visual/material, a reconceptualisation that was particularly attentive to the embodied, material, and political nature of representations. It links ongoing concerns in the Social Studies of Science literature that foreground questions of how matter comes to matter in politics or, in Latour's (2004) words, how things can be made public and an object orientated democracy realised with the performative metaphysics elaborated by Barad, which might provide an ontological basis for such 'politics with things'.

Below we introduce our case study, as well as a recently created forum, the Loweswater Care Project, that came into being precisely in order to make things present and acknowledge their agency. The focus of the chapter though is not on this forum *per se* but on a specific kind of scientific practice supported by it – echosounding. Echosounding in Loweswater lake involved three hydro acoustic surveys during which multiple material intra-actions took place. These intra-actions (a term we describe in full below) rendered visible millions of transparent 'phantom midge' larvae and brought them into the political setting of the Loweswater Care Project *visually* in form of echograms. In the chapter we are keen to highlight the processes, practices, and contingencies involved in this 'making visible' and to explore the implications of making these an explicit part of the forum's deliberations in practical, ethical and political terms.

The chapter thus simultaneously grapples with concerns that have occupied scholars working on issues relating to the 'visual turn' and the 'material turn', and it does so in a spirit of diffraction (Haraway, 1992, 1997; Barad, 2003, 2007). This means that it thinks the 'social' and the 'scientific' together:

Like the diffraction patterns illuminating the indefinite nature of boundaries – displaying shadows in 'light' regions and bright spots in 'dark' regions – the relation of the social and the scientific is a relation of 'exteriority within'. This is not a static relationality but a doing – the enactment of boundaries – that always entails constitutive exclusions and therefore requisite questions of accountability (Barad, 2003).

Such a refocusing takes us away from representationalism and its historical preoccupations with questions of truth and accuracy, such as whether scientific knowledge represents independently existing entities and how faithfully it does so, or whether language and its concepts and categories do justice to the external referents they invoke. This is achieved through a performative understanding of the visual and the material where both are considered as intra-actively participating in the world's becoming.

The Politics of Things: Reflections on the Materiality of Representational Practices

In the epigraph, Doreen Massey challenges us to think critically about the widespread notion of 'landscape' as something stable and unchanging. Instead, she proposes the more dynamic understanding of landscape as *provocation*, a provocation that stems from the ever-changing materialities of specific landscapes. In the midst of such change, she observes, we can often experience a sense of loss. This can engender fears of what might happen, a sense of an unknown future that is threatening and difficult, as expressed in the second quotation coming from an interview with a Loweswater farmer and his wife conducted in March 2008. Such fears tend to become embodied in conservative political strategies of preservation. Massey points to the English Lake District in which Loweswater, the location of the intra-activities described in this chapter, lies, in particular as an area often represented as unchanging even though its' very mountains, as she notes, have historically been on the move. Indeed few English landscapes provoke such strong feelings *against* change in England as does the Lake District. Many people imagine it as a place of seemingly unspoilt nature, open valleys, dramatic mountain scenery, traditional farming practices, and ancient crafts traditions. They hold on to these images for fear of losing a reality that in fact, in such un-ambivalent form, never existed there in the first place (Grove-White et al., 1994; Urry, 1995; Tolia-Kelly, 2007). Feeding these images are countless poetical and pictorial representations to go back several centuries: poems by the Romantic poets, especially William Wordsworth; paintings by John Constable and William Turner; photographs, postcards, calendars and, more recently, movies (including a Hollywood take on the famous children's author Beatrix Potter, [1] one of the leading advocates of the preservation of certain material characteristics (particularly farmsteads) found in the region in her day).

While images of the landscape of the Lake District as unchanging are thus upheld by a plethora of stabilizing representational practices, the materialities and processes that co-mingle to form that landscape continually undermine them: this is one interpretation of Massey's conception of landscape as *provocation*. Another interpretation leads us more deeply into questions of politics and ontology, such as the ones addressed in Jane Bennett's (2004) work on an ecology of matter that highlights the interconnectedness of the human and the nonhuman through the vitality and energy of matter, a thing-power potent enough to provoke and change the course of human action, or Michel Callon's notion of material 'spill-overs' which implies that the harder we try to frame and conceptually 'bind' the world around us, the greater the failure of our bifurcation efforts will be in practice (Callon, 1998). These reconceptualisations of ontology are necessary for a politics *with* things that encounters scientific visual representations as actants. They emphasise the positive political potential of a more lively and culturally engaged ontological understanding and political engagement. To expand on this point, Bennett's 'thing power materialism [...] focuses on *energetic forces* that course through humans and cultures without being exhausted by them. [...] Its political potential resides in its ability to induce a greater sense of interconnectedness between humanity and nonhumanity'. Thing power materialism recognises 'the agential powers of natural and artifactual things' – a sense of agency also found in Karen Barad's (2007) posthumanist agential realism and expressed in her notion of intra-action discussed in more detail below – agential powers that once acknowledged, so Bennett suggests, can foster a 'greater awareness of the dense web of their connections with each other and with human bodies, and, finally, a more cautious, intelligent approach to our interventions in that ecology' (Bennett, 2004: 3349).

Change and provocation are relevant to our discussion in this chapter which concerns Loweswater, a small hamlet of houses and farmsteads scattered around one of the smallest lakes in the North Western Lake District. Loweswater lies within the Lake District National Park and has been designated by the Authority as a 'Quiet Area' (North West Regional Development Agency 2004, 2005). This means that change – particularly to the material fabric of the place and in relation to intrusive economic development – is discouraged and/or prevented through planning powers. The Quiet Area designation has helped maintain a familiar sense of 'time-stood-still' in Loweswater, where the look and feel of the landscape, its basic infrastructures (dry-stone walls, small country roads) and its dwellings have been deliberately preserved over decades (Davies and Clarke, 2010). However, beneath this unchanging surface or veneer, Loweswater is inevitably changing in many ways, and many social, economic, and cultural shifts have occurred in

---

1   Miss Potter (2006), Director: Chris Noonan, Writer: Richard Maltby Jr., Cast: Renee Zellweger, Ewan McGregor and Emily Watson, Producer: The Weinstein Company.

recent years. What we focus on here, though, is a change in the ecological fabric of the place that includes the soil, land, water, and all the organisms and processes belonging to these media, but which is expressed and becomes visible through the water of Loweswater lake.

Studies of water quality in Loweswater note a gradual deterioration over the past sixty years (Bennion and Winchester, 2010). This deterioration is not generally detectable by eye, but the flourishing of blue-green algae, made visible by ever more frequent occurring algal 'blooms' on the lake. This is a stark indicator and reminder not only of deteriorating water quality (caused by the element phosphorus moving from land to lake water) but that wider ecological relations in the valley have changed in perhaps nefarious ways since at least the mid-1950s. This changing and mobile materiality of Loweswater provokes, in Massey's (2006) sense, the 'ways of seeing' (Cosgrove, 1992) Loweswater that have been promoted through the Quiet Area designation of Loweswater by the LDNPA – a visualization and representation that has acted as a powerful cultural, political, and ideological device (see Davis and Clarke, 2010). The recurring blooms also undermine the romanticized visual representations of the Lake District discussed earlier in this chapter more generally, representations that 'endure through different media, creating sanctioned ways of viewing' (Witcher, et al., 2010). However, relations between the visual/representational and the material are never straightforward. Representations, as visual theorists remind us, are not disembodied from the worlds they represent (Pink, 2003; Rose, 2003; Witcher, et al., 2010). Rather, they are co-produced and co-constructed through continually emergent, intra-active associational relations between actants perceived as 'visual' and 'material'.

We suggest below that the potentially toxic algal blooms, recurring at unpredictable intervals in Loweswater, are an example of a material spill-over of 'idyllic' Loweswater. Having witnessed the way in which these blooms have constituted a powerful provocation – both to local residents and businesses and to the agencies with responsibility for environmental quality in the area – we see them as putting into question dearly held notions of Loweswater as a pristine, unchanging environment and fuelling fears expressed above of Loweswater as 'dying'. The blooms are a *material* provocation, sometimes visible, sometimes not, but always potentially recurrent and often unexpected (e.g. during the winter months). We suggest in what follows that the blue-green algae that constitute such blooms can be seen as 'matter' that is of concern in the sense of Latour's notion of 'matters of concern' that have the capability to draw together a concerned public (Latour, 2004, 2005; see also Marres, 2007). Furthermore, we show how thinking about algal blooms in this way has influenced the 'public' that has gathered in their name. We describe this public below – a collective of residents, researchers, environmental agencies, and a growing array of nonhuman things that, since 2007, has carried out research and worked together within a new organisation of their own making, the Loweswater Care Project,[2] to try and understand what causes these blooms and to see how to work together as a collective that may, or may not, want to act, to 'do something' about them.

From 'Matter of Fact' to 'Matter of Concern'

For Latour, conventional ecological politics is fundamentally flawed because of its reliance on Science as the predominant means by which nature's materialities are understood in politics.[3] Nature, he suggests, is rendered as 'matters of fact' via Science, and this, he argues, tends to downplay or even supplant the importance of political processes. This view of Science, Latour suggests, has its origins in Enlightenment thinking and underpins what he perceives of as the project of Modernity (Latour, 2004). He urges us to rethink this view and put centre-stage the ways in which materialities as 'matters of concern' have the power to call forth assemblies and new collectives that meet in their name: Parliaments of Things whose political nature and force is acknowledged and done justice. 'Things' are no longer understood here as the passive, powerless stuff that is controlled, moulded, changed, and shaped by humans at their will. Through the reconceptualisation of 'things' – including scientific visual representations, as we will show below – as matters able to call forth a public, their political nature as actants that co-produce worlds and intra-act with humans in complex ways is brought to the fore (Latour and Weibel, 2005; Latour, 2005).

---

2   The Loweswater Care Project (LCP) was formed as part of a Rural Economy and Land-Use Program (RELU) funded research project entitled 'Understanding and Acting in Loweswater: a Community Approach to Catchment Management', and was initiated by an interdisciplinary group of scientists based at Lancaster University. The authors would like to acknowledge the contributions of Lisa Norton and Stephen Maberly from the Centre for Ecology and Hydrology, Nigel Watson from the Geography Department, and Ken Bell a Loweswater farmer and part-time researcher to the findings presented in this chapter. The project begun in 2007 and ran until December 2010. The LCP is likely to continue as a locally-led participatory action group beyond 2010. Further information about the project and its rationale can be found at <www.lancaster.ac.uk/fass/projects/Loweswater>.

3   Latour, *The Politics of Nature: How to Bring the Sciences Into Democracy* p. 9 asks his readers to 'disassociate the sciences – in the plural and in small letters – from Science – in the singular and capitalised. I ask readers to acknowledge that discourse on Science has no direct relation to the life of the sciences, that the problem of knowledge is posed quite differently, depending on whether one is brandishing Science or clinging to the twists and turns of the sciences as they are developed...'. He goes on to define Science as the politicisation of the sciences through epistemology in order to render ordinary political life impotent through the threat of an incontestable nature' (p. 10) and contrasts this to the sciences whose task is to create collective propositions with which to constitute the world. See glossary entry on p. 249 of Latour 2004. Latour, *The Politics of Nature: How to Bring the Sciences Into Democracy.*

We rehearse Latour's arguments on issues of nature and science here because they have formed an important part of the research and research collaborations we have begun to sketch out above. They also bring additional insight to the complex ways in which materialities and visualities intra-act, particularly in visual form, and how they play a performative role in political processes which can have a significant reconfiguring effect on how humans intra-act with nonhuman entities.

Initially, concern about blue-green algae in Loweswater gave rise to the foundation in 2002 by local farmers of the Loweswater Improvement Project. This was exclusively concerned with farming practices and their potential impact on Loweswater. One main reason for the coming together of these farmers was a situation of conflict between local farmers and the owners of the lake, the National Trust, as well as other agencies concerned with the deteriorating water quality of Loweswater. At this time, farmers felt that they were being held to blame for the state of the lake. Blame was a recurring theme in conversations with Loweswater farmers and in the meetings that were held as part of the forum that constitutes the basis of this chapter. Following recent elaborations in visual/material theory it could be said that this issue of blame stems from the distinct embodied encounters that different actors had with Loweswater. If we agree with critical visual theorists that images of Loweswater '"script" ways of seeing and being in the landscape' that can lead to very different encounters and expectations of this place (Witcher, et al., 2010), it is not surprising that conflicts of vision ensued. Following Grasseni (2004), these different visions can furthermore be viewed as 'skilled visions' acquired through education and training 'in a relationship of apprenticeship and within an ecology of practice'. Such skilled visions became an even more important factor in the discussions that ensued in the expanded group that formed in Loweswater around the matter of concern of blue-green algae in 2007, the Loweswater Care Project, hereafter LCP. By the mid-2000s, algal blooms had become a matter of sufficient concern amongst local residents, farmers, small businesses, social, and natural scientists and environmental agencies to support the creation of a new, expanded organisation. With the involvement of scientists from Lancaster University and a research institute who were interested in experimenting with Latour's ideas of 'doing politics with things' in practice, several open meetings were held in Loweswater with residents, farmers, and agency representatives to see whether there was enough interest among this collective to participate in a forum that declared its experimental nature from the outset. Between 2007 and the end of 2010, the LCP came together fifteen times in the name of blue-green algae in Loweswater Village Hall, once every two months and for an entire evening (5.30–9.00 p.m.). It encompassed a multitude of materialities, visualities, and actants including farmers, local people, agency representative, natural- and social scientists, and others concerned about the lake. Many of the ideas that underpin the LCP came from Latour's *Politics of Nature*, particularly the notion that 'matters' are powerful forces in calling forth 'new collectives', and that such collectives of people and things should consider how nature is known and represented in affairs that are essentially political ('matters of concern', rather than 'matters of fact').

In the LCP, participants, including the authors of this chapter, experimented with these ideas of Latour to see whether it was possible to create a new collective in which understandings and representations of nature were approached in a fresh and challenging way, and which would ensure that such understandings of nature were:

> not self evident; where knowledge and expertise has to be debated; where it is accepted that uncertainty is the main condition humans are in (rather than a condition of having knowledge); where what is important is the creation of connections between people and things; and where doubt and questioning is extended to our own representations ...' (Latour, 2004).

This was no mean feat because of the unconventional nature of the approach. Participants of the LCP, the authors and fellow researchers tried to keep 'watch' over these principles and encouraged the different aspects of the short statement above to be acted upon. It involved time, effort, and especially at the beginning, episodes of conflict. However, gradually over time our distinct skilled visions began to converge and as a group we managed to achieve the conditions necessary for vigorous and frequent debate. We became comfortable talking to each other about what was not known, or was uncertain, as well as what was known; we applauded and supported the creation of new connections between people and things (which could involve carrying out a small piece of research that might connect the algal blooms in Loweswater to issues at other registers and scales – such as European water policies, or changes in farming subsidies and practices); and we ensured that the many new representations that were created through debate, investigation and research were exposed to any doubts and questions that participants had, no matter how seemingly trivial, controversial or otherwise they were. In this chapter we shall dwell no further on the methodological issues raised by trying to do politics with things collectively in the LCP and how individual participants dealt with the challenges that working in this way posed for them; rather we want to move on now to the implications of refocusing participatory efforts on matters of concern rather than matters of fact and we will do this through the example of the scientific practice of echosounding and the particular scientific visual representation it co-produced of a hitherto invisible Loweswater actant, *Chaoborus*.

To make more explicit what a political, epistemological and ontological shift from matters of fact to matters of concern entails, let us just briefly beforehand try and illustrate using Loweswater's blue-green algae, scientifically known as cyanobacteria, as an example. If cyanobacteria had been treated by the LCP in a purely *matter-of-fact* way, we might simply have noted the following about them:

- Blue-green algae are a form of bacteria: hence their name, cyano*bacteria*;
- Cyanobacteria are photosynthetic organisms and belong to an ancient family of organisms thought to have existed for at least 3.5 billion years;
- Water is central to their life cycle;

- They produce food by photosynthesis, using chlorophyll;
- They absorb carbon dioxide from the atmosphere and use solar energy to transform it into complex energy-containing sugars. While doing so, they release oxygen. To do this, each cell needs to be exposed to bright sunlight;
- Cyanobacteria live in sheet-like films thin enough to allow each cell to have access to both water and sunlight. To stay together, they secret a slime;
- Cyanobacteria form complex, layered colonies. Underneath the surface layer is another layer of photosynthetic bacteria that absorb sunlight at wavelengths where the cyanobacteria are transparent. Because these are poisoned by oxygen, they are called 'anaerobic'; the first mat of cyanobacteria acts as a protective shield. Further down, additional underlying layers that can be millimetres or centimetres deep contain other forms of anaerobic bacteria. As they do not receive sunlight and do not conduct photosynthesis, they feed on dead photosynthetic bacteria left behind by the gliding of the live ones toward the sun;
- Cyanobacteria thus have a unique gift of buoyancy: they are able to store tiny volumes of gas which enables them to remain near the surface of a water-body to feed on solar energy. Once their carbohydrate store is full, they become heavy and sink to the bottom where they feed on nitrogen and phosphorus. Before their carbohydrate store depletes, they rise to the surface again to absorb sunlight again;
- Under certain conditions – such as in nutrient-rich still waters – they can reproduce rapidly to form algal blooms;
- Some cyanobacteria produce extremely potent toxins when dying or damaged and livestock and dog deaths have been reported in many countries, including England, after animals came into contact with algal scum;
- When and where toxins are released is difficult to determine;
- Algal blooms contribute to what is called eutrophication which can cause living organisms in a lake to die.

Although these matters-of-fact about cyanobacteria are fascinating and valuable in that they point us to the complex material intra-actions in which they are embroiled, they tell us nothing about the political nature of cyanobacteria and the ways in which they co-configure the worlds they form part of, in this case, Loweswater. This political nature comes to the fore in their relations with others, including human beings. Consider, e.g. how blue-green algae as actants became a *matter of concern* that called forth a public (the LCP), and you can see that the presentation of an a-political cyanobacteria is a fiction. Cyanobacteria as political actants play a central role in the performance of the LCP where they challenge and intra-act with embodied visions and ways-of-seeing Loweswater: they feature in local stories and gossip; inspire certain actions and prevent others; allow certain things to be said but not others. Indeed, without them, the LCP would not have come into existence.

Similarly, in the LCP, scientific representations of cyanobacteria in the form of images, texts, photographs, and statistics among other things are not simply taken as objective representations or 'mirrors of nature'. Rather, they too are considered as actants that intra-act with other materialities in complex ways, including humans and their scientific apparatuses and practices. These practices and the many contingencies that bring various actants to human consciousness are an important part of their presence in the LCP and constituted an essential part of debate and questioning. Thus the LCP supported two kinds of representative practice: first through the injunction to let connections and representations proliferate, it sanctioned many different kinds (scientific, lay, amateur, professional,and collective) of investigation, and any kind of connection to the blue-green algae that participants considered worth pursuing. The 'results' of these connections and investigations were brought before the collective for debate and they came in many forms – as photos, as graphs, as physical objects, as texts, as arguments. But second, the LCP encouraged that all these representations, and particularly the ways in which they had been produced, be scrutinised, doubted and questioned. For example, on the occasion where several different scientists brought together their results concerning the flow of phosphorus in the valley and represented these in the form of a Power Point presentation to LCP participants, the latter were stimulated to hear the question posed as to whether a certain field in the valley had been included in data gathering, and if not, why not? Thus not only 'results', but the way in which the investigation had been framed and conducted were legitimate questions and challenges for debate. In other words, the LCP was *both* a forum for making invisible elements and processes visible, and for bringing materialitites into the collective, *as well as* a forum in which the very practice of rendering things visible and present through (scientific) visual representational practices was laid open to critical scrutiny. This second emphasis, on making explicit the practiced, performed work of 'naturalcultural practices', as Barad calls them, emphasises the activities that bring representations to bear and firmly locates them within matter: that 'dynamic intra-active becoming that never sits still' (Barad, 2007: 170).

Before looking in detail how one particular actant emerged through scientific investigations into cyanobacteria at Loweswater, we want to briefly examine Barad's notion of intra-action here. Similar to Latour's endeavour to shift attention from matters of fact to matters of concern, Barad through her posthumanist performative ontology attempts to overcome representationalism in the sense of representations as mirrors of nature and the realist assumption that the material world exists independently out there waiting to be discovered by human beings. For her, humans are part and parcel of matter and thus from the beginning 'inside' the project; intra-action and the dynamism of matter for her is 'generative not merely in the sense of bringing new things into the world but in the sense of bringing forth new worlds, or engaging in an ongoing reconfiguring of the world' (Barad, 2007: 170). For Barad, intra-action, *signifies the mutual constitution of entangled agencies* ... in contrast ... 'interaction'... assumes that there are separate individual agencies that precede their interaction ... intra-action recognises that distinct agencies to don't precede, but rather emerge through, their intra-action. (Barad, 2007: 33).

Barad's idea of agential realism, like Latour's ideas of matters of concern that call forth a collective, has been a further influence on the way that the LCP has understood its own research, discussions and other activities. It has provided a constant reminder that all scientific and investigative practices are co-constituting the objects they produce, and that there are many contingencies in the ways in which new knowledge of the world we live in emerges. Below, we present a particular example of scientific intra-actions that helped such a new world emerge and find its way into the LCP.

Making the Invisible Visible through Sound: Loweswater's Echo-sounding Surveys

In the LCP, one of the questions that scientists, local people, and agencies considered worthy of attention, given anecdotal evidence and given that water quality was known to have been deteriorating for several years, was this: 'how many, and what kind of fish still remain in the lake at Loweswater?' This simple question triggered a scientific investigation that involved, among other activities, the carrying out of three hydro-acoustic echo-sounding surveys. These, in turn, brought forth a new 'thing' that came to constitute a matter of concern in its own right in the collective: the *Chaoborus* or phantom midge larva. The story of the discovery of *Chaoborus* and the complex intra-actions that led to its coming forth into the collective are told below in form of a first-person account. Ian J. Winfield, a scientist with the Centre for Ecology & Hydrology at Lancaster and head of the surveys carried out in the summers of 2007, 2008, and 2009, remembers the complex material and representational intra-actions that led *Chaoborus* to become part of the emerging story of the LCP.

Echo-sounding Loweswater and Finding the Unexpected

When I was invited to become involved in this project to conduct a study of the fish community of Loweswater, I was particularly pleased because at that time I knew almost nothing about the lake. Much of my other research on lake fish populations has involved long-term studies at sites which are visited repeatedly, often for many years, during the course of an investigation. The novel blank canvas offered by Loweswater was thus very attractive.

The scientific study of lake fish populations is notoriously difficult. While many animal ecologists may argue the same for their own specialist areas, I think that this is particularly true for fish for three major reasons. Firstly, fish are difficult to sample in a scientific manner. That is to say, although it is easy to go out and catch a few fish with a rod and line, it is much more difficult to take a more extensive and unbiased sample of the fish living in a lake. Generations of fish and fisheries scientists have had to invest huge amounts of effort in developing what were basic hunting skills and tools into a range of scientific sampling techniques. Secondly, fish are relatively large vertebrates which engender significant ethical issues to their study, as well as making them of considerable interest to fishery owners and other members of the public. Both of these features mean that the study of fish can be a complicated and political business. Thirdly, and most relevant to the present discussion, the fish ecologist has the distinct difficulty of usually being unable to see his or her study species. As human beings, we rely greatly on our sense of sight in order to make sense of our world. Scientists studying animals such as birds, deer or rabbits enjoy the great benefit of a 'visual check' on their study species, even though they will also typically employ a range of other sampling techniques in their investigations. In contrast, scientists studying fresh waters are rarely able to see anything of their subjects. Low underwater visibility usually precludes the use of direct (through diving or snorkelling) or indirect (through underwater video or still cameras) visual observations as a significant part of our information-gathering activities. As a keen teenage angler, I remember sitting on the banks of numerous lakes and ponds wishing that we had a machine which would allow us to see into their waters.

My first thoughts on how best to investigate the fish community of Loweswater thus took into account the above generic difficulties characteristic of fish studies, added to which were the real world limitations of the resources available for this work in terms of specialist equipment and staff time and the added complexity that fish studies require a number of permissions. I decided that our basic approach would be to use a combination of scientific survey gill netting and echo-sounding. The former technique has been used for decades and is now developed into a European standard describing an agreed design of net and an agreed way in which several such nets are to be set in a lake overnight to take a scientific sample of the fish community. In this way, biological specimens can be obtained to provide information on the fish species present, their individual conditions, their growth rates, and their diets. However, gill nets have a major limitation in that they almost invariably kill the fish that they sample. While this is not an issue for commercial fisheries on lakes, it is more of a problem on lakes used for recreational fishing purposes like Loweswater, and so I tend to use them as little as possible. Rather than set large numbers of gill nets, I prefer to use the much more sophisticated and completely harmless technique of echo-sounding. This process involves a hardware and software system generating short pulses of very high frequency sound, beyond the hearing ranges of humans and fish, out into the water and then 'listening' for returning echoes. The latter are detected as pressure waves in the water and converted immediately into digitised electronic signals which are taken back into the software component of the system and displayed as marks on an 'echogram'. Echoes are produced by any objects with a density different to that of the water through which the sound waves are passing. In lakes, these objects are typically the lake bottom, plants, small animals and fish. Because the speed of sound through water is extremely fast at around 1.5 km per second, this entire process of transmit–listen–record is repeated a number of times during each second of the sampling. During a survey, all of these echoes together with their precise timings and location information from a global positioning system reliant on overhead satellites are recorded in data files within the echo-sounding system. In essence, for each target encountered, be it the lake bottom or living organisms within the water column, from the time taken for the echo to return the technique gives us the target's distance and direction from the boat (and thus its depth) and from the echo's strength an estimate of its 'acoustic size', which can subsequently be converted into its physical size with increasing strength of echo indicating an increasing physical size of the target. In many ways, such echo-sounding systems give me the 'underwater sight machine' that I so longed for in my youth, although they operate using not light waves but sound waves. In fact, light actually hinders echo-sounding surveys because in order to avoid their predators during the day fish typically hide in physical structure or assemble together in just a few large shoals. As a result, daytime echo-sounding surveys often record very few fish and it is only after nightfall that fish move out from their refuges and spread more evenly throughout the water, at which time they can be more effectively surveyed.

Having decided on the above general approach and obtained all of the appropriate permissions, the next step in our study was to formulate a field plan specifically tailored to Loweswater. I decided on site locations for a minimum of netting to cover the inshore, offshore surface, and offshore bottom habitats of the lake's three dimensions. In addition, I plotted a course over the lake which would allow our echo-sounding to record data from twelve transects spaced systematically across the lake. These location data were then uploaded to a handheld global positioning system to facilitate on subsequent navigation on the lake itself, both during daylight and during the night.

Our first action was for my two colleagues to follow the plotted route to undertake the first ever echo-sounding survey at the lake.[4] During this time, I stayed on the shore to allow myself to take a step back and to think through everything one last time to make sure that I had not overlooked potential complications which were not evident during the initial planning stage, but may have become apparent during our first real look at the lake. All was as expected and our first echo-sounding survey was completed without incident. My colleagues returned to shore and I got my first look at the collected daytime echo-sounding data. Full analysis of such data requires a laboratory examination using specialist computer programmes, but here at the lakeside, and indeed in real-time on the boat as the

4   We thank our colleagues Janice Fletcher and Ben James for their assistance in the field and laboratory. We are also grateful to the Environment Agency, the Lake District National Park and the National Trust for granting the appropriate permissions necessary for our field activities. We are particularly indebted to Mark Astley of the National Trust for his hospitality and logistical support at Loweswater. Finally, we thank Simon Pawley of the Freshwater Biological Association for permission to use his photograph of a Chaobrous larva.

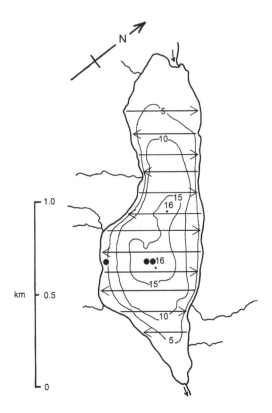

Figure 7.1   **The twelve transects (arrows showing direction of travel across the lake) followed during the echo-sounding surveys of Loweswater with the locations of the three survey gill nets shown as closed circles. Redrawn in part from Ramsbottom (1976)**

Source: Ramsbottom, A E *Depth Charts of the Cumbrian Lakes*. Freshwater Biological Association Scientific Publication No. 33 (1976).

survey progresses, the data can be seen displayed visually as an echogram on the laptop of the echo-sounding system. Such echograms are produced for each transect and can be visualised as a cross-section of the lake, with the lake's surface running along the top of the screen and its bottom being visible as a very strong and continuous echo in the lower part of the echogram. Any echoes originating from organisms appear as small marks between the surface and the undulating bottom of the lake metres below. As explained above, the system in fact records a wealth of information for each echo and the detailed examination of these data must be undertaken later in the laboratory, but here in the field a visual inspection of their patterns can be used to check that the system has performed correctly and to gain an overall impression of the lake's habitats and its inhabitants. The system had indeed performed perfectly, but the results obtained were remarkable and somewhat unexpected.

Given reports that anglers visiting Loweswater had found fish increasingly difficult to catch in recent years, I had expected to detect very few targets in the lake's water column. However, at depths below about 12 metres the water contained huge numbers of acoustically weak (and thus physically small) targets. This was somewhat surprising because I knew that for many weeks over the summer, Loweswater held little oxygen in its deepest parts which were thus inhospitable to fish. In addition to this unexpected abundance, the echogram also showed that these small targets were rather uniformly distributed in this lower part of the water column, i.e. they were not grouped into daytime shoals as may be expected if they were small fish. This mystery deepened further when my colleagues and I went out again in our boat and took measurements of the temperature and oxygen conditions at regular intervals from the lake's surface to its deepest point. These measurements confirmed that the typical situation of the summertime Loweswater persisted on that specific date, with the lower, cooler water of the lake almost devoid of oxygen. So, these small targets were not only very abundant, they were largely restricted to a part of the lake almost devoid of oxygen. From my experiences elsewhere and knowledge of the scientific literature, I knew that at least some of these acoustic targets were of the physical size of small fish, but most of them seemed to be too small to be attributable to even the smallest fish likely to be in the lake.

Figure 7.2   Photo of boat on the lake. First echo-sounding survey of Loweswater, 2007. Part of the echo-sounding system is visible over the port side of the boat

Source: © Ian Winfield

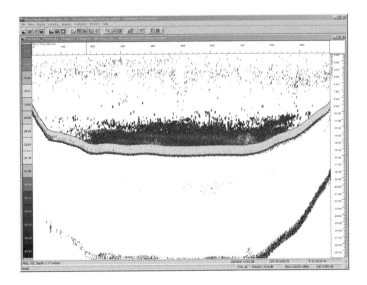

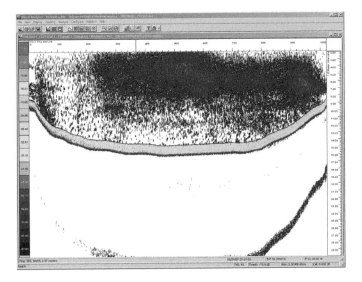

Figure 7.3   Example echograms recorded during the daytime (upper
              figure) and night-time (lower figure) echo-sounding surveys of
              Loweswater. In both cases, the bottom of the lake can be seen
              as the thick line running across the bottom of the echogram.
              Numerous small targets can be seen within the water column
              in both echograms, but note their contrasting vertical
              distributions

Source: © Centre for Ecology and Hydrology, Lancaster

Even though these daytime echo-sounding findings had been a great surprise, we continued with our initial plan of carrying out a gill-netting survey and repeating the echo-sounding survey after dark. My colleagues prepared three survey gill nets for setting in the lake, with one set with ropes and weights appropriate for its deployment in the shallow water of the inshore area where fish may be expected to take refuge, one prepared to be set floating nearing the surface of the open water where many fish may be expected to feed, and one prepared to be set at depth in amongst the multitude of small acoustic targets. We then set these nets in the late afternoon, to be lifted during the next morning after they had entangled their catch. After darkness fell, two of us (this time myself included) went out once more on to the lake and repeated the echo-sounding survey, following precisely the same route as taken in daylight a few hours previously. This time, I was able to see each echogram building up echo by echo as we transversed the lake. Huge numbers of small targets were again recorded, but now their vertical distributions were almost a mirror image of their daytime patterns and largely restricted to the top 8 metres. I went to bed in a nearby hotel puzzled but looking forward to the recovery of the gill nets first thing on the next day.

The next morning all three of us got into our boat and went out to recover the three gill nets. On the basis of talking to anglers and other local people, we had expected to catch a few brown trout (*Salmo trutta*), perch (*Perca fluviatilis*), perhaps pike (*Esox lucius*) and minnow (*Phoxinus phoxinus*). In fact, we caught just brown trout and perch, although subsequent netting has indeed added pike and minnow to our species list for the lake. We quickly returned to the shore were we removed all fish from the nets, carefully keeping the catch of each net separate in labelled plastic bags which were then frozen. At a later date back in the laboratory, the fish of each bag were thawed and then subjected to detailed examination including the measurement of their individual lengths and weights and the determination of their individual ages.

The distribution of the brown trout and perch amongst our three nets was particularly interesting and had both expected and unexpected dimensions. The few brown trout that we recorded ranged in length from 153 to 166 mm and were contained exclusively in the net which had been set near the lake surface out in the open water. This is a typical finding for this species in lakes. In contrast, all of the perch, which ranged in length from 79 to 367 mm, were found only in the net set in the shallow water of the lake's edge. I had expected to catch most perch in this area, but even so I was surprised not to find even a single perch offshore. The lack of any fish in the net set in the deepest part of the lake is somewhat unusual, but here at Loweswater the finding could be readily explained by the very low oxygen levels recorded at this location during the previous day and which typically persist throughout the summer.

To recap, our daytime echo-sounding survey had found highly numerous but small targets offshore in deep, oxygen-poor areas of the lake which at night-time migrated up towards its surface. Over the same time period, our netting had found large numbers of perch inshore and a few brown trout in the offshore surface waters, but no fish were found in the deep areas where small acoustic targets were common during the day. Furthermore, although these small acoustic targets approached the lake's surface at night, the only fish caught there were a few relatively large brown trout which could not have produced the observed echoes.

This state of affairs presented a puzzle unlike anything we had seen in similar surveys at several tens of other lakes. The small acoustic targets, which were so numerous and which were evidently animate in nature because they showed such a marked vertical migration over twenty-four hours, were initially a mystery. While they were of a physical size which could have been very young perch, such fish would usually not be present in the deepest part of a lake during the summer and certainly not in the local absence of plentiful oxygen as observed at Loweswater. Although initially mystified by these tiny objects, I knew that certain small invertebrates could in theory be detected by our echo-sounding system although we had never previously encountered them in our surveys elsewhere. One type of invertebrate in particular seemed to be a potential candidate as the originator of the kinds of echoes observed in our survey, i.e. the phantom midge larvae *Chaoborus*. This animal is the aquatic life stage of a small fly and reaches a length of about 10 mm (Figure 7.4); it is also known to be able to tolerate very low oxygen availability. Its body is highly transparent, hence the name 'phantom', but it also contains two pairs of small gas sacs which have two important consequences in the present context. Firstly, they enable *Chaoborus* to control its buoyancy and thus confer on it an ability to make rapid vertical migrations over several metres up and down the water columns of lakes. Secondly, they give this invertebrate a significant

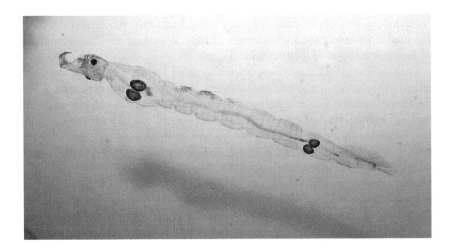

Figure 7.4   A *Chaoborus* larva of approximately 10 mm in length. Paired gas sacs are visible near each end of the elongated body

Source: © Simon Pawley (Freshwater Biological Association)

acoustic signature of magnitude rarely seen in aquatic invertebrates. A combination of a detailed analysis of these remarkable echoes and reference to the scientific literature on *Chaoborus* acoustics confirmed that these tiny animals were indeed the most probable sources of the echoes observed in Loweswater. At a later date, we returned to Loweswater with a fine net capable of being opened and closed remotely, thus allowing us to sample specific parts of the water column for Chaoborus. As expected, we found the highest numbers of these animals at the depths of the lake where the small echoes had also been recorded at their highest abundance.

Rendering Visible, Encountering the Material

Part of the philosophical commitment of researchers, local residents, and agencies within the LCP was to set aside received wisdom, unquestioned assumptions and ideas, and to create a forum in which we could explore, connect, and question. To allow, as Latour above suggests, representations and connections to proliferate, at the same time as allowing fierce doubt and questioning of those representations to take place. From the researchers' point of view this was a commitment that made room for a practical, experimental encounter with an ecology of matter (Bennet, 2004), with an intra-active, emergent world (Barad, 2003, 2007), a commitment that took seriously spill-overs as effects of prior framing, and as issues that deserved further research in their own right. Connections between different kinds of things were necessary to make visible the invisible, to make present what had hitherto not been materially present in any political fora.

But things, we found, could be politically elusive. The blue-green algae of Loweswater are ephemeral and unpredictable. Sometimes they appear, causing a stir in their surroundings; then they disappear, sink to the bottom of the lake, and become invisible for long periods of time almost to be forgotten. Similarly, phosphorus, – which in Greek means 'light-bearer' because the element emits a faint glow when exposed to oxygen – which nourishes the blue-green algae and controls its abundance, is notoriously difficult to trace and virtually impossible to see. It too disappears to the bottom of the lake, binds onto the sediment, and 'recycles' within the lake. Phosphorus's main sources are farm fertilisers, animal manures, slurry tanks, human domestic waste waters, and leaky septic tanks. But to bring this elusive, fluid and sometimes invisible element to witness takes time, knowledge, experience, patience, money, effort, and labour.

Ian J. Winfield's account above provides a window into the complex procedures, multiple technologies and apparatuses, diverse knowledges and experiences, and the vast amount of resources needed to render 'things' visible. And yet unless they can be brought to our attention in some way, we do not know that they are there and even if we sense their presence, we are uncertain as to what they might be. In the case of *Chaoborus*, their discovery depended on many contingencies. Only a few decades ago, the technology of echo-sounding for researching lakes was not available; as Ian J. Winfield puts it above, [for me] the 'echo-sounding systems give me the "underwater sight machine" that I so longed for in my youth', and the fact that these systems use sound waves instead of light waves hardly matters. And yet, it is incredible to think what is involved in making *Chaoborus* visible through 'sound' – trained people, lakes, boats, cables, satellites, sounds, software, algorithms, screens, printers, paper ... the list could go on. And then, of course, there is knowledge and experience – the skilled vision referred to earlier – which requires ability, schooling, and practice. Winfield often refers to the 'mystery', the 'puzzle', presented by the graphs produced of Loweswater, and through his account we get a glimpse of the reasoning involved in working out what these dots on the graphs might mean, and what they are. As a result of his 'experience elsewhere and knowledge of the scientific literature', or in other words, his practiced and professional eye, he is able to identify the thousands of dots on the graph as phantom midge larvae, an initial guess later confirmed by further investigations in the field – or better, the lake! Phantom midge – what an apt name for this elusive actor!

And yet even though *Chaoborus* were introduced into the political arena of the LCP through scientific visual representations, in many ways they were part of it long before they were made visually present. In spite of their absence from human knowledge they were present as something going on in the lake, something that may have contributed to diminishing fish stocks and intra-acted with blue-green algae. Something that was sensed and suspected, observed as a disruptive force but not visually detected except in the noticing of their observed effects. The creation of a visual echogram featuring these larvae forcefully brought home the realisation that there might be many more unknowns that lurk in the lake – a reality that we think we know and yet one that eludes us in so many ways.

Materialities and visualities, are thus always part made, co-made, through the multiple collectives they co-produce in their intra-actions. There are presences and absences, surprises and discoveries, knowns and unknowns, and once they are acknowledged, they humble our certainty of thinking that 'we' are the ones in charge. Working with and through things and questioning how they intra-act with us in forums such as the LCP can make this abundantly clear: even our final representations that were brought into the forum were only ever a particular 'cut' (Barad, 2007), a particular 'take' on reality and a partial and fragile one at that.

Some Final Remarks on the Loweswater Care Project

So how has the LCP, over the past two and a half years, succeeded in constructing and moulding itself and its human and nonhuman inhabitants? As we hope will be clear by now, the LCP was never a forum designed to discuss pre-established things. But it was a forum in which participants could creatively work with ideas, tools, local knowledge, different forms of expertise and many other things (including living organisms, water, computers, budgets, policies etc.) to make new things visible and present for participants. It was also a forum in which a wide variety of connections with a diverse range of registers and scales were carefully followed up. These connections would later be brought back into the collective in the form of new representations where they would be scrutinised, questioned, argued about, and used to make more connections, questions or statements. The LCP, in other words, was intensely generative – of ideas, images, texts, graphs, models, etc. It was a shaping force in the construction of new visibilities and in the making present of ephemeral elements, organisms, and relationships.

But the things that came to be present in the LCP had, for the authors at least, a special and valuable quality: they were things tightly accompanied by their own history of being made, as part of the collective. They bore and displayed the imprint of recent collective discussions, of decisions taken, of the care and craft involved in bringing them to light, of the contingencies of their making, of the new questions and uncertainties that they generated. Whenever discussing new ideas, findings or claims, *how such things had come to be* seemed to be as important within the LCP as the ideas or claims in themselves, and so these things were not disembodied facts, but embodied and shaped realities, all taking their own place in an increasingly populated collective. Within a relatively short period of time, the LCP successfully forged a rich sensibility to the making present of things. So, when Ian J. Winfield wraps up his observation on his own fish surveys as follows, (as he also recalled verbally in an LCP meeting one evening):

> What had been expected to be a routine fish survey had produced some very unexpected results which took a little detective work to explain. My interpretation of the assembled observations is that the persistently low oxygen levels of the deepest areas of Loweswater offer Chaoborus a refuge from the attentions of visually-feeding perch which would otherwise predate them down to very low population abundance. At night, when the feeding efficiency of the perch is reduced by the low light levels, the Chaoborus are free to migrate higher in the water column and themselves feed on the smaller zooplankton of the open water, which in turn have been feeding on the small plants (algae) of the system. Although subject to further investigation, it appears that in contrast to the other lakes of this part of the U.K. where small fish are the major predator of the zooplankton, in Loweswater the peculiar environmental conditions interact to result in Chaoborus assuming this role.

He may interpret this as making one small contribution to the overall 'picture' emerging about conditions in Loweswater Lake. However, the other two authors of this chapter (Tsouvalis and Waterton) would interpret this kind of statement and its reception within the LCP further. The echo-sounding surveys, carried out as part of the LCP's urge to know more about the fish in the lake in fact opened up a whole new sensitivity among participants. Through these surveys participants began to grasp: the idea that the scientist–nature encounter is one that conventionally involves a large amount of craft and experience, but also a large amount of guesswork, contingency, and unknowns. Scientists do not always know what they are making present/visible and mostly work in a terrain of vast uncertainty. Participants also took on board a much more complex understanding of the entire aquatic ecology of the lake, and the idea that subtle changes in the biological food chain within the lake may have a profound influence on the fate of blue-green algae. Lastly, participants also began to understand more deeply the specific material particularities of this lake relative to other 'similar' lakes, and of the significant impacts that even small differences can have. In sum, participants' eyes were opened to the intricate liveliness of the context in which blue-green algae can survive and thrive and to the limits of our human efforts to fully understand this context.

We began this chapter reflecting on the way in which the landscape around Loweswater is often assumed to be unchanging and stable, and is deliberately preserved that way through the material cultures of tourism and the visual representational practices circulated by the Lake District National Park Authority and other institutions that offer this landscape to the nation as a place of authenticity, pleasure, and quiet enjoyment. We were keen, however, to stress that no matter how durable this idea is, provocations in the landscape are occurring and, in the case of Loweswater's blue-green algae, are present enough to become a matter of concern. This provocation, as Massey puts it, initially seemed like a disruption and a menace, creating fear of a dead lake, even a dead community. However, what we would like to suggest is that this provocation, taken up by the LCP as an opportunity to do nature–culture politics in new ways, has actually created the conditions for a rich engagement and set of mediations within and between the human and nonhuman worlds. The LCP, informed in part by the theories of science studies and feminist materialism was partly an experiment to see if an institution could create visualities and welcome in materialities whose history and contingency could also be simultaneously acknowledged. In some senses, therefore, it was an experiment in creating power/empowerment, at the same time as refusing that such created power goes unquestioned. This seems to have made space for a new kind of sensibility, and a kind of reflexive acting that perhaps brings with it new ethics and an emerging new sense of responsibility. Bennet has written about the prospect of a 'more cautious, intelligent approach' to our interventions in ecology arising from an acknowledgement of the agential powers of natural and artifactual things. Only time will tell whether the considerable achievements of LCP participants working together to do politics with things will yield such desirable results.

References

Barad, K. (2003), 'Posthumanist Performativity: Towards an Understanding of How Matter Comes to Matter', *Signs: Journal of Women in Culture and Society*, 28:3, 801–831.

Barad, K. (2007), *Meeting the Universe Halfway – Quantum Physics and the Entanglement of Matter and Meaning* (Durham and London: Duke University Press).

Bennett, J. (2004), 'The Force of Things. Steps Towards and Ecology of Matter', *Political Theory*, June, 367.

Bennion, H. and Winchester, A. (2010), 'Linking Historical Land Use Change with Palaelolimnological Records of Nutrient Change in Loweswater, Cumbria', Lancaster University, Report produced for the Loweswater Care Project (LCP), <http://www.lancaster.ac.uk/fass/projects/loweswater/research.htm>.

Callon, M. (1998), 'Essay on Framing and Overflowing', in M. Callon (ed.), *The Laws of the Market* (Oxford: Blackwell Publishers/The Sociological Review).

Cosgrove, D. (1984), *Social Formation and Symbolic Landscape* (New Jersey: Barnes and Noble Books).

Davies, D. and Clarke, E. (2010), 'Community and Culture – Tourism in a Quiet Valley', Lancaster University, Report produced for the Loweswater Care Project (LCP), <http://www.lancaster.ac.uk/fass/ projects/loweswater/research.htm>.

Grasseni, C. (2004), 'Skilled Vision. An Apprenticeship in Breeding Aesthetics', *Social Anthropology*, 12:1, 41–55.

Grove-White R, Darrall, J., Macnaghten, P., Clark, G. and Urry, J. (1994), *Leisure Landscapes* (Lancaster: CSEC, Lancaster University).

Haraway, D. (1992), 'The Promise of Monsters: A Regenerative Politics for Inappropriate/d Others', in L. Gossberg, C. Nelson and P. Treichler (eds), *Cultural Studies* (New York: Routledge), 295–337.

Haraway, D. (1997), *Modest_Witness@Second_Millennium.FemaleMan_Meets_ OncoMouse: Feminism and Technoscience* (New York: Routledge).

Latour, B. (2004), *The Politics of Nature: How to Bring the Sciences Into Democracy* (Cambridge: Harvard University Press).

Latour, B. (2005), 'From Realpolitik to Dingpolitik or How to Make Things Public', in B. Latour and P. Weibel (eds), *Making Things Public – Atmospheres of Democracy* (Cambridge and Karlsruhe: MIT Press and ZKM).

Marres, N. (2005), 'Issues Spark a Public into Being: A Key but Often Forgotten Point of the Lippmann-Dewey Debate', in B. Latour and P. Weibel (eds), *Making Things Public: Atmospheres of Democracy* (Cambridge and Karlsruhe: MIT Press and ZKM).

Marres, N. (2007), 'The Issues Deserve More Credit: Pragmatist Contributions to the Study of Public Involvement in Controversy', *Social Studies of Science*, 37:5, 759–780.

Massey, D. (2006), 'Landscape as Provocation. Reflections on Moving Mountains', *Journal of Material Culture*, 11:1/2, 33–48.

Miss Potter (2006), Director: Chris Noonan, Writer: Richard Maltby Jr., Cast: Renee Zellweger, Ewan McGregor and Emily Watson, Producer: The Weinstein Company.

North West Regional Development Agency (2004), 'Lake District Economic Futures: The Way Forward'. A Final Stage 2 Report. Regeneris Consulting.

Pink, S. (2003), 'Interdisciplinary Agendas in Visual Research: Re-situating Visual Anthropology', *Visual Studies*, 18:2, 179–192.

Rose, G. (2003), 'On the Need to Ask How, Exactly, is Geography "Visual"', *Antipode*, 35:2, 212–221.

Tolia-Kelly, D. (2007), 'Fear in Paradise: The Affective Registers of the English Lake District Landscape Re-visited', *Senses and Society*, 2:3, 329–352.

Urry, J. (1995), *Consuming Places* (London: Routledge).

Witcher, R., Tolia-Kelly, D. and Hingley, R. (2010), 'Archaeologies of Landscape. Excavating the Materialities of Hadrian's Wall', *Journal of Material Culture*, 15:1, 105–128.

Chapter 8
Materialising Vision:
Performing a High-rise View

Jane M. Jacobs, Stephen Cairns and Ignaz Strebel

Introduction

Red Road high-rise housing estate in Glasgow was designed and constructed between 1964 and 1968 and comprises two 26–28-storey slab blocks and six 32-storey point blocks. At the time of construction it was the tallest residential development in Europe. Red Road was a product of the post-war, state-led, programmes of mass housing provision in Britain. It was a materialisation of a specific modernising vision for cities and city life, one that joined the progressive ethos of state-led welfarism with a modernist architectural aesthetic (Figure 8.1).

Given its size and symbolic station it is unsurprising that Red Road played a key visual role in the opening shots of a forward-looking 1971 film entitled *Glasgow 1980*, commissioned by the Glasgow City Corporation. Following the opening title we are taken on a sweeping aerial view of the just completed Red Road estate.

Figure 8.1   The Red Road estate, Glasgow
Source: Authors

Directed by Oscar Marzaroli and filmed by Martin Singleton *Glasgow 1980* offered a mid-term showcase of the Corporation's progress with respect to its twenty-year redevelopment plan. As such it is part documentary, part booster advertising. *Glasgow 1980* folds together modernist urban aesthetics and infrastructures, the vision of social democracy, and film – that unique visual medium of modernity. *Glasgow 1980* delivers to us content (high-rises, motorways, get up and go music, a rallying voice over) that operates as an illustration of a place being re-formed in the name of an urban vision. The aerial view of Red Road that features in *Glasgow 1980* also inevitably speaks to the reverberation between visuality and materiality. It is a visualisation of a building technology that communicates that the estate is manifest, and has a specific and spectacular shape and size. That manifest housing development, by being incorporated into this film, takes up symbolic purpose and is able to stand for the successful materialisation of a modernising mass housing vision. Through such films we better grasp, as Lebas (2007: 50) puts it, 'the part which imagery of the city played in its construction'. Yet in every other way this representational entrance to Red Road elides its materiality: through the frame of *Glasgow 1980* the estate can only ever be an object that has been subjected by the scopic logics of modernisation boosterism. One strategy for moving beyond this prison is to think more materially about the making of this specific Red Road representation. We might, for example, interrogate the technologies, practices and uses of the sweeping aerial view in urban and architectural documentary traditions. We might, alternatively, ask more about the circumstances of the film's production and set it into a wider political and film-making and film reception context, as Lebas (2007) has. That is, one strategy for saying more about the way materality and visuality are entangled in the city, it to access activities that are beyond the image but which contribute to or are effects of image-making.

This chapter both relies upon and wishes to extend such strategies for entering into the rich relationship between visualities and materialities. We see Red Road as an object of visualisation, but also as a technology that supports and depends upon what Denis Cosgrove (2008: 5) called 'vision in the sense of active seeing'. This chapter explores this proposition by way of an investigation of the 'active seeing' that happens in and around high-rise windows, and specifically the windows of Red Road. The window is a complex assemblage of things (material and immaterial): design specifications, material components (glass, jambs, frames), building and safety regulations, mechanisms such as hinges and locks, and decorative artefacts such as curtains. In unison with human users it becomes a purposeful aperture between the interior and exterior of a building: letting in light, ventilating, offering views (both out and in), as well as metaphysical opportunities such as 'allowing our own spirit to penetrate outwards' (Markus 1967: 103). It also offers other, un-programmed opportunities: an opening to jump from or through which to throw rubbish. The window is the technology that manipulates the distribution between inside and outside in terms of light, air, noise, comfort, and safety.

We want to link the windows of Red Road and the viewing practices they afford to the visualities of other orders, and including: avant garde architectural visions, urban visions, the visualisations of housing and building science, electronic surveillance, and the glance. The chapter enters into the practices of urban visioning of which the modernist high-rise was a part, but it also addresses the role of views and viewing technologies (such as windows and cctvs) in that architecture, as well as the ways in which such visioning technologies are lived with. The chapter explores how these varied visual orders and practices bear directly on the materiality of a building: bringing it into being, holding it in place (with varying degrees of success) and even contributing to its demise. We also wish to show that vision does not exhaust the sensorium of the window. To see the window simply through the lens of the eye and sight overlooks the extent to which windows operate in other ways to order (and disorder) architecture.

In telling this story we attempt to bring in the many participants (human and non-human, material and immaterial) that play a part in the interacting practical orders that produces the visions and visualities of Red Road. Interactions such as these are always productive, although that productivity may incline either towards 'success' or 'failure'. Like all building technologies, a window (and specifically a window-in-action) may perform what it was programmed to do or it may err towards that to which it was not programmed: it may become an opening to jump from or through which to throw rubbish, activities that Hand and Shove (2007: 81) call 'anti-programming'. It may also, through technological failure or redundancy, become de- or re-programmed (a window that does not open, or a window with broken glass that lets in too much of the outside). In this sense performativity and (building) performance are linked. Wanda Orlikowski (2005: 185) has suggested that we might even think of something she calls 'material performativity' in which 'human agency is always materially performed just as material performances are always enacted by human agency', wherein both are 'temporally emergent in practice'.

Modernism's Visions

The architectural imagination that gave rise to the high-rise as a mass housing solution was motivated by the potentials – economic, formal, social, spatial – of new materials such as steel, innovative construction technologies such as rapid system-building, new sciences about health and environment, and innovative mechanisms such as passenger lifts and integrated garbage handling systems. Le Corbusier's famous aphorism – 'a house is a machine for living in' – makes perfectly clear that this avant-garde architectural vision depended upon embracing technology: efficiency, rationality, standardisation, mass-production. This technological orientation was given expression in mass high-rise housing in his *La Ville Contemporaine* (1922) (reproduced in his *Urbanisme* [*The City of Tomorrow*] (1987 [1925]). As geographer David Pinder (2005) reminds us in his account of Le Corbusier's city vision, this dogmatic, technologised and formalist project was linked to a passionate utopian hope in the power of architecture to re-shape society for the good.

State agencies in the UK came to adopt, develop, and implement specific policies based upon the architectural avant-garde's 'ethos of optimism about technology' and its faith in the universal applicability of mass production building technologies. This vision came to 'link ... high flats with technological advance' as 'a leitmotif' of post-war housing reconstruction (McCutcheon, cited in Dunleavy 1981: 59). In the UK, the optimism that saw the rapid and large-scale construction of high-rise housing in the 1960s was quickly tempered by a range of technical and social problems manifest therein. As one housing historian has commented, '[h]igh-flat building was the most extreme and conspicuous form of mass housing provision [and] ... has since become one of the most widely proclaimed (if unstudied) "failures"' of public policy in this field' (Dunleavy 1981: 3). Social scientific accounts of high-rise living often sought to test the social effects of living in this novel housing type and there is an extensive scholarship in post-occupancy studies that measure and chart resident satisfaction with high-rise living. One such study, directly relevant to Red Road was Pearl Jephcott's study published in 1971 under the title *Homes in High Flats: Human Problems Involved in Multi-storey Housing.*

In such studies high-rise housing serves as a laboratory for testing and observing problems residents might be experiencing with this novel technology for living: problems with lifts, communal circulation systems, garbage disposal and washing, and, of course, living with height. A good number of these studies evidence a technological determinism, in that it is the high-rise form (and its novel technologies) that is called upon to explain the quality of life of life of residents within. In contexts like North America and Britain, it was this kind of social science that did the work of verifying the emergent problems of high-rise housing, and which, in these contexts, translated a housing solution into a housing failure, most spectacularly evidenced by the demolition of Pruitt Igoe in St. Louis.

In Le Corbusier's original vision of the high-rise city, the window/view had a very specific place. This may be illustrated by a sketch he did in *La Ville radieuse* (1967 (1933)) (Figure 8.2).

Here a high-rise vision is stripped of its formal features and represented simply as a set of gathered-together services (water, electricity, gas, telephone). Atop this unclad high-rise block is a single apartment, occupied by a tiny figure and a disconnected, distended eye looking outwards. Architectural critic Beatriz Colomina (1994) has argued that this eye serves to communicate Le Corbusier's commitment to the window as an aperture for light and seeing (see also Friedberg 2006: 123). This commitment to light and seeing was set against the nineteenth-century domestic interior. That earlier interior was turned inward from the world outside as if a refuge (Perrot, Ariès and Duby 1990). As Walter Benjamin (1999: 221 and 216) put it, the transparency and openness of Le Corbusier's modernism 'put an end to dwelling in the old sense' and to its 'nihilistic cosiness'. The modernist window shuns curtains and other 'hampering objects' and 'throws the subject towards the periphery of the house' (Colomina 1994: 283).

Much has been written about how Le Corbusier broke from the tradition of the 'vertical window' and developed the expansive 'horizontal window'. This transformation did not simply alter the parameters by which light or air were let into the interior. As Friedberg (2006: 126) notes, the horizontal window also encased 'a panorama that ... brought more of the outside in'. Modernism's window restructured the relationship between occupant and exterior. The vertical window positions the occupant/viewer as central to the framing and, by taking in foreground, horizon and sky, replicates perspectival depth (Colomina 1994: 128–30, see also Reichlin 1984). In contrast, Colomina argues, modernism's horizontal window replicates the vision of the 'camera, the mechanical eye' in which (especially with the movie camera) there is no central, centred viewing position (Colomina 1994: 133). Furnished with the horizontal window the apartment

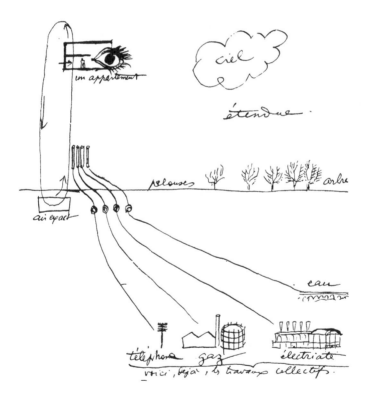

Figure 8.2   Le Corbusier's unclad high-rise apartment block, showing the enlarged eye and the diminished human occupant

Source: Le Corbusier (1967 (1933): 21). © FLC/DACS, 2008)

becomes a 'viewing mechanism' in which the views taken in are choreographed by the occupant. As Friedberg (2006: 128) notes, the frame of the window and its glass expanse become only one element in the inhabitor's ... mobile view'. The relationship of window to view has been taken up by Isenstadt (2007) in his study of the emergence of the large-scale, plate glass suburban window in America. His scholarship shows that the window's relationship to viewing landscape has been both elaborated through design (bigger and more carefully placed windows) and consolidated by real estate advertising into a domestic asset called 'a view'.

We have already noted how Le Corbusier's vision of a clean and aerated high-rise city was the distant template for many of Britain's state sponsored, post-war high-rise housing developments, including the likes of Red Road. Yet what was built in the name of modernism in such mass housing programmes usually dispensed with key design elements. Nonetheless, when the first and tallest of the Red Road flats was formally opened on the 28 October 1966, the window was enrolled into the publicity efforts that marked the materialising of this housing vision. Front page news of the day was a picture of William Ross, the then Secretary of State for Scotland, and his wife, looking out of a top storey window at the construction site (Figure 8.3).

Figure 8.3   William Ross, then Secretary of State for Scotland, and his wife, at a Red Road window on the occasion of the estate's official opening, 28 October 1966

Source: Image courtesy of *The Herald & Evening Times* picture archive

At this celebrated moment a high-rise window and its view was mobilised to speak for the wider project of post-war mass housing provision. This was not an uncommon iconography of the inaugural moment of high-rise living. For example, an early image from a Pruitt-Igoe promotional brochure pictures what we must assume to be a recently housed family gazing out from a large window across the urban landscape. In such images, the high-rise window and its view represent the hope of the new life offered by high-rise housing. Yet the viewer/window/ view assemblage captured in the Red Road image is distinctly arranged. It is not panoramic or moving. It is a still image that shows people poised, close in to the window, pane ajar, looking and pointing. This inaugural image of a Red Road view both confirms the view as an asset of living high, but also illustrates some of the peculiarities of how a Red Road view has to be achieved. The Red Road view is made not by passively standing in front of a glass expanse (as in some of the iconic images of architecture's modernist window), nor even (as Colomina suggested) through some newly mobile viewing practice. It is made, so it would seem, by getting up close to the window, opening it and looking out.

Seeing Truth

We have already mentioned the various social science studies of high-rise living and building performance. These studies would often use the window as one of their truth spots: verifying the claims made about high-rise housing and their views, with the reality of what was built (Gieryn 2006). For example, the 1971 study of high-rise resident satisfaction in Glasgow, Jephcot's *Homes in High Flats*, paused specifically at the window of 'Mrs ... , of Wyndford'. Careless design meant that while the window captured a view, this view could only be seen from an impossible point near to the ceiling. A seated 'Mrs ... , of Wynford' could not enjoy the view at all for it was blocked by a poorly placed eye level balustrade.

Building science responded to such problems by starting to articulate systematically and scientifically what comprised a 'successful window' and 'view satisfaction'. One such building scientist was Thomas Markus, better known to many of us for his later book *Buildings and Power* (1993). In his 1967 paper 'The Function of Windows' he set out to specify the minimum and optimum performance specifications for windows in high-rise buildings, based on a detailed study of the Robertson high-rise office block in Bristol. Part of his concern was with conditions such as light penetration, a factor that was pertinent to the deep spaces of open plan multi-storey offices, and he diagrammed systematically the changing penetration of light through the windows across the course of a day. He was also concerned with the matter of the view, which he saw to be central to the optimal performance of windows. For Markus, satisfactory views needed to have a certain compositional character: a 'horizontal stratification' incorporating a layer of ground, a layer of city or landscape, and a layer of sky. A 'successful window', he argued, must have a fenestration that facilitates this framing (the right placement, frames and fixtures that do not block the dividing lines between these layers). Here we might imagine that Markus is simply resistant to the abstraction delivered by the modernist window, and nostalgic for the pre-modernist vertical window and its perspectival framing of the external world. But Markus also argues that a successful window needs 'dynamics': that is, it should offer a good amount of changed views as a result of changed viewing positions of the observer. Markus advised that none of this be left to chance or to speculation by the designing architects. He advised, instead, that architects use a photo-based montage methodology for assessing view content and variations for occupants.

View Events

Today, Red Road estate is an emblematic example of the dramatic slide from utopian vision to dystopian reality that has marked so many post-war high-rise modernist social housing programmes in Britain. It has suffered a history of dis-investment, residualisation and stigmatisation. In 2005 the Glasgow Housing Association made a determination that Red Road was no longer a 'sustainable' or 'viable' social housing development and announced a £60m redevelopment strategy, the first stage of which is to be the demolition of the block with the address 213/183/153 Petershill Drive. It is to the windows and views of this soon-to-be-demolished high-rise building that we now turn.

We wish to use these windows to open up current thinking about the relationship between the modernist window and the view. In doing so we take rather literally Cosgrove's invocation of 'active seeing' and draw on materials generated in a wider project on the many afterlives of the middling modernist residential high-rise form (<http://www.ace.ed.ac.uk/highrise/>). As part of that study we produced a building biography of a soon-to-be-demolished block in Red Road. That work involved interviewing housing officials and activists, doing video-ed walking and talking tours of flats with residents (what we dubbed a 'Show us your home' exercise) and conducting a video-based, workplace study of the concierge service for the building. As we talked and walked through the flat or its communal spaces one of the things we asked our guides was to 'show us your view'. In what follows we dub these instances of active seeing as 'view events': sequences of action wherein a building occupant shows us and explains to us their view. As will be clear, this solicitation did not simply produce a description of what was outside and seen: what might be thought of as the view in landscape terms. Rather, it revealed how a view is a product of a set of complex and contingent interactions between that outside world that is looked upon, the framing technology of the window, and the viewers.

The high-rise view events we recorded involved as many as four viewers (the resident or worker, the interviewer, the camera person, and the camera). These, then, are not normal view events and we do not claim that our video data captures anything like a naturalised scene of an occupant taking in a view. We and our

cameras are not merely capturing these viewing performances we are, alo
the residents/workers and the building, both making them and configuring t
a variant of what Douglas Macbeth refers to as the 'praxeology of seeing
camera' (1999: 151). At various times the camera and its operator is quite act
the viewing events analyzed: being invited closer in by the resident/guide, h
doors and windows opened for it to look through, being shown where to poi
this sense, the residents and workers of Red Road were, as much as us, direc
of the sequences shot.

Also central to these Red Road views are the design and materials of I
Road's windows, their surrounds and the many other objects that come
attach themselves to the technology and space of the window. When built, t
prefabricated, single glazed, steel-frame windows of Red Road were at the forefro
of window technology. Furthermore, being steel framed they were stitched int
local initiatives that sought to link the building of the 'all steel' Red Road to effort
to revive the fortunes of the local steel industry which was, post-war, suffering a
slump (Jacobs, Cairns and Strebel 2007). Although they are, relatively speaking,
more horizontal than vertical in dimension they are also, relative to wall area,
modest apertures with the windowsill located at a standard height (they are not
the floor to ceiling windows often associated with modernist architecture). Also,
these windows open and close like normal domestic windows, although because
they are windows in a tall building they are equipped with devices that inhibit
this action for safety purposes. These additional technologies ensured that under
normal circumstances the window opened only a small way, such that people (and
especially children) and things could not fall out. Finally, the Red Road windows
were not supported by a collectively delivered window cleaning service, such
as one might see on commercial high-rise buildings. Red Road windows were
domestic windows and were to be cleaned by residents themselves. To facilitate
this they featured relatively novel tilt-and-turn hinge technology for ease of
cleaning thereby ensuring that residents did not place themselves at risk in the
interests of a clean window.

All these attributes shape the ways in which Red Road views are conducted in
practice. For example, the size of the windows of Red Road, mean that they were
a long way from any modernist ideals of a transparent glass screen. As an ageing
technology that is imprecisely lived with, these windows interrupt the smooth
alignment of window and view in many ways. They have curtains, hold ornaments,
carry stickers, get dirty and cracked, and even, in some cases, have smoke damage.
They also carried lots of talk. Often enough that talk was about the view to be seen
('my view here is lovely', 'it is just too beautiful', 'see doun there, look at that,
in'that beautiful? 'see at night ... its really terrific', 'It's not a very nice view ...
except perhaps with the empty space. Here at least got that ... '). Sociologies of
technology have shown that talk always goes with technology in action. This is
often more so with failing technologies which solicit commentaries of various kinds
to mediate their deficient performances, including inquiries, apologies, excuses,
explanations, and complaints (Latour 1988, Law 2003). For example, one elderly

resident when asked to show us her view took us immediately to the window and started pointing, but her first directive to us was not for us to 'look at the view' but to 'see' (that is, notice) but not 'look' at (that is, not focus on) the window itself: 'See this doun here ... see the damp on the windae, don't look hen, no. It's really mank'.

During the course of showing us her view, this aged resident repeatedly apologised for the state of her window. This state was a consequence, in part, of the age of the windows, which for reasons we will discuss later were still the building's original ones. It was also a consequence of the timing of our visit. Because this resident was about to leave the soon-to-be-demolished Red Road, she explained, she had not bothered to clean the windows. This then led her to explain how, even under normal circumstances, maintaining the cleanliness of this heavy and aged steel-framed, tilt-and-turn window was no small job (Figure 8.4).

In a sense this resident's apologies and explanations were themselves a kind of discursive maintenance – doing the work of repairing a window whose performance as a viewing technology had been compromised.

In yet another recorded view event, we see again the complexity of taking in a Red Road view, especially if one is also showing that view to some researchers and a camera. As in the photograph of the inaugural Red Road view, and in the one we have just been shown, this view event required the window be opened. In this case the act of opening the window to show us the view solicited not a lament about the state of the window, but a complex emotional terrain (cf Kraftl and Adey 2008, Rose et al. 2010). This resident is rapturous about here view: 'My view here is lovely, it's just so beautiful the view, really, outside'. Yet we also glimpse that this high and expansive view may have first been experienced as a kind of sublime, generating fear and anxiety as much as awe and wonder (Figure 8.5).

As is clear from the resident commentary included in Figure 8.5, when this resident first lived with this window and its view she was not comfortable. She may have looked at the view from the safety of the room, but she was frightened of the window that afforded her that view: she would not approach it, nor open it (she accidentally calls it a door, as if she might have stepped through it), nor look down through the opened window. Coming to learn to happily co-exist with her high window and the view that it affords has been a process of domesticating the sublime through routinely handling the window technology – approaching it, cleaning it, dressing it, opening it, poking one's head out of it. Despite all this work, the 'wonderful' view that the window affords retains some anxiety for this resident. Although she confidently negotiates the window and now proudly accesses 'my view', this relationship is not extended to her 10-year old child. As the resident explained to us:

> I have to close them [the windows] all the time because of the child. And NASS [National Asylum Support Service] come and check the windows shutters [catches] every time because they are scared ... that if the window is not working properly, the shutters are not closed properly, and you have a small child it is dangerous (Red Road Resident, Show Us Your Home Transcript 2005).

Res: No you got to get up into it [the window], son, and you will see it all and you will see all, all up there, you see all the hills [open out the window and puts head out] There. See all the hills.

Int: Oh yes

Res: I look out this windae and see the hills. See them all?

Int: Yeah.

Res: But see at night you'd appreciate this better. You would see all the lights [hand sweeps across indicating distant panoramic view - **still a**]. And you can see doun there, look at that, in' that beautiful? [still b] And you see…I'll bring [addressing the camera operator and moving out of the way to allow him in to window space - **still c and d**] Look doun there, look doun there son [addressing the road] see at night they are all…[breaks off - **still e**] its really terrific…

Figure 8.4   Red Road resident explaining how to clean her window

Res: At first I wouldn't open the door [sic] and look, open the window and look down [hand gesture goes down - **still a**]. It was so scary. But now I look around and it is just too beautiful. You can see the view from that side [hands gesture to outside]

Figure 8.5   Red Road resident explaining her first experience of her view

For the child the window is encrusted with rules (Wood and Beck 1994). Some of these are local and familial and entail the parents not allowing the child to open the window. Others are technological and include the fact that, unlike many other flats we visited, the child safety catches in this flat were all in working order. They were so because this flat was the property of the National Asylum Seeker Service and these residents were asylum seekers awaiting determination of their refugee status. NASS routinely check on the state of the housing of those for whom they are responsible. With their residency so enframed, this window was not only a worry for the mother near-to-hand, but also for a distant agency such that the window's safety lock was delegated the responsibility of keeping a child safe.

In another Red Road flat, accessing 'the view' was more difficult. After we requested the resident show us her view, she simply stayed on her sofa with window curtains drawn and began describing to us what might be seen from the window if one were looking through it. Only after specific prompting did this resident go to the window, pull back the curtain and show us her view. Furthermore, unlike all the others view events we have discussed thus far, this resident did not feel the immediate need to open the window to show us her view. What produced the delays and interruptions to (the showing of) the view in this instance? Was it a result of the resident feeling as if it were not worth showing and looking? It may well be that, in this resident's opinion, what was outside did not conform to some preconceived notion of a view. Were we asking her to show us something she did not think she had, even though her flat was on the top storey of Red Road? When we finally did get to the window, the view once again had to be seen through the literal and figurative haze created by the window technology and the talk that surrounded it. This window, like others we saw and saw through, was damaged and old. Specifically the glass had permanent smoke damage from a chip-pan fire some years ago. As the resident explained of her view once we were at the window, ' ... well you can't really see this. Again [as she puts her hand up to the window to gesture to the view], with the fire, this window smoked up'. In Red Road, the showing of a view happens through an active seeing, in which the window's part is far more (and sometimes far less) than a technology of transparency.

Broken Windows

We have seen to date a sequence of Red Road view events, each of which reminds us of the active seeing that is required to take in a view, and nowhere more so than in a context where the viewing technology of the window is broken or compromised. Through these examples we have tried to show something of the imperfect practice of modernism's windows. In this section we wish to extend our investigation of the relationship between materiality and visuality in another direction. Specifically, we want to look at how an artefact like a broken window can re-shape not only how a high-rise development like Red Road is seen, but also how seeing happens in high-rise living.

When high-rise living environments came to be associated with various social pathologies (crime and deviancy among them) the window was often drawn in to the emergent social science as an indicator of all that was wrong with high-rises. For example, in positing a wider thesis on the importance of architectural design in determining quality of life, William L. Yancey drew specifically on the example of the Pruitt-Igoe Housing Project. In one 1971 article, reprinted in Kaplan and Kaplan's *Humanscape* (1978), he recounts his experience of entering Pruitt-Igoe and notes that many of the collective spaces were 'lined with broken windows' (Yancey 1978: 299). For Yancey such 'physical danger and deterioration' was but a 'reflection of the more pressing human dangers' (299). Illustrating his article is an image of a broken Pruitt-Igoe window.

We can see in such studies (many of which were aligned with a form of environmental determinism, sometimes simplistic, sometimes more sophisticated) the origins of a wider quasi-scientific theory about crime and urban environments known as the 'broken window theory'. Brought into prominence in the 1970s by George Kelling, this 'theory' is directed towards promoting a more localised (walking-the-beat) form of policing, justified on the grounds that indicators of neighbourhood disrepair (such as a broken window) foster criminality. As Kelling and Wilson (1982: 2) put it 'one unrepaired window is a signal that no one cares [and] such an area is vulnerable to criminal invasion'.

While the broken window was positioned as an indicator of disorder in the high-rise, the matter of proper seeing was deemed to be an important part of solving the problems of high density, high-rise estates. As an aperture, the window allowed for all-important lines of sight a complex built environment with many collective spaces containing corners, corridors and stairs, layouts that could produce so-called 'blind spaces'. This was evidently so in Oscar Newman's 1972 'defensible space' thesis, which was built upon a detailed study of crime and environment in New York high-rise housing projects (Newman, 1972). And as part of that study Newman proposed that where windows no longer offered lines of sight for residents, alternative technologies of surveillance could be used. One such technology was what he dubbed the 'acoustic window', which was an aperture that would allow for the aural monitoring of public space. He also outlined ideas for the use of cameras and microphones in lifts such that any unruliness might be broadcast via loudspeakers to corridors and apartments, or ferried to on-site security offices.

During the 1980s and early 1990s technologies of security such as these became an important part of the housing policy initiatives that attempted to make high-rise estates safer and more pleasant places in which to live. Red Road estate, for example, underwent a version of such an upgrading. The entrance and exit points of the buildings were adjusted spatially, secured electronically and stitched into a staffed concierge service. The entrance and exits, common spaces and lifts were wired up to an electronic cctv surveillance system, managed by the concierge from their central monitoring station. In our workplace ethnography of the concierge system at Red Road we saw such these alternative viewing systems, both human and non-human, in operation.

In the first instance the work of security and maintenance surveillance at Red Road was done through an embodied practice, a routine activity known as the block check (Jacobs and Cairns 2010, Strebel 2011). This is a system of seeing and checking the public spaces of the building and involves the concierge doing a routine walk of the building's common spaces, starting at the top of the building and walking down common stairs and landings. This is work that depends on the eye as a tool of verification, routines of walking and looking through which the current state of the building is checked against a normative vision of how the building should be (in terms of look, safety, use). The work is about constant movement of concierge eye-to-building, building-to-eye. On the routine block check little attention is given to the views afforded by Red Road's windows. For the most part they pass without notice. The concierge is primarily, although not exclusively, interested in the fabric and function of the building itself and the well-being of residents. In a context such as this, when the building being cared for is coming to an end, the nature of this maintenance work shifts. There is, for example, little need to record and act on many of the building failings seen on a routine check, as there might be in a building with a longer future.

The block check is a once a day event. The more sustained and distributed visual work done by the Red Road concierge service depends on interacting with the cctv surveillance system. On a usual shift the concierge will spend a good part of the day located in the concierge station. They pass time chatting as they keep an eye on their bank of cctv screens. Such electronic surveillance systems are inserted into fabric of the building in order to allow spaces that are out of view to be in view – to produce new levels of spatial coordination and visual integration. That transparency is produced in order to ward off deviant and damaging behaviour (it is supposed to be both a technology of deterrence as well as a technology of detection). This is a manifestation of the now well-documented gaze of securitization, modernity's paranoid gaze.

In Red Road these embodied and virtual surveillance routines were stitched into a wider set of localised checking and monitoring practices, understood by the concierge and residents alike as being as much about maintaining Red Road, as securing it (see also Heath and Luff 2000). As an instrument of work, the Red Road visualistion systems rearranged the geography of Red Road so that the concierge could undertake their work of securing and maintaining beyond where they were. The system pulls the distant space of the lift into the concierge service station, and (by way of a lift speaker system) delivers the authority of the concierge into the lift when needed. The cctv and the concierge work to join up the building. In their albeit rather improvised and incomplete centre of calculation in the concierge station, cctv technology and concierge work together to monitor the spaces of Red Road and determine, based on what they see, whether intervention is needed (for example, a visit to a landing, a voice in to the lift, a call for the police). Together these visual technologies and human viewing practices play their part in keeping Red Road in place as a satisfactory housing solution.

Wind

It is the ambivalent moral order of visual surveillance systems such as that of Red Road, which forms the basis of Andrea Arnold's 2006 noir thriller, entitled *Red Road*. Filmed in and around the soon-to-be-demolished Red Road this film was also drawn to the domestic widows of Red Road and their views. The main protagonist of the film, Jackie, works in City Eye Control station watching and over a dozen television monitors that survey the streets of Glasgow. Her job is to alert the police if she sees any suspicious activity. One day while viewing the area around Red Road she sees a familiar figure, a recently released criminal we come to know as Clyde. We never really know what this man did to destroy Jackie's family, but once Jackie sees him she vengefully stalks him. Her efforts to get close to Clyde takes her inevitably to Red Road, where she links up with some of his friends, later visiting them in a typical Red Road flat. There Clyde's friend excitedly and unexpectedly asks Jackie and his girlfriend if they 'want to feel the wind?'. He leads the women to a tilt-and-turn Red Road window, and then throws it wide open. it The women gasp as their breadth is taken away by the shock and strength of the wind that streams in. Recovered and tentatively seeing rather than just feeling what this open Red Road window affords, the man lifts his girlfriend up and terrifyingly pretends to push her out the window. This thrilling, tension-filled cinematic Red Road window is a long way from le Corbusier's modernist equivalent. It is not only (or even at all) about the view. It is an aperture that opens out to a chill Scottish wind, facilitates a threatening joke, and interacts with bodies that are actively engaged in an 'anti-programming' of the window by doing very un-view-like things (Hand and Shove 2007: 81).

The English term 'window' derives from the Middle English vindauga, eye of the wind ('vindr' – wind and 'auga' – eye). Thus far, this chapter, like much cultural analysis of the window to precede it, has privileged the 'eye': the 'ordering of the gaze', the en-framing of views, the performed view (Harris and Ruggles 2007: 27). In the final section we would like to reactivate the relationship between window and wind. The work a window does as a building technology extends beyond that of vision. The window is at core an opening between inside and outside. In the modern window glass is central to this achievement, its transparency keeping the 'outside out and at the same time bringing it in' (Friedberg 2006: 113). But glass is not alone in the making of the modern window. It is joined by other materials (steel, wood, aluminum, pvc, fabric), working in other ways (as frames, casings, jambs, hinges, locks, hinges, decoration, insulation), to manage the opening of which they are a part. Furthermore, an opening is always relational. The work a window opening does is relative to the walls around it (window to wall ratios and window placement). In unison with a human user (and in some 'smart' buildings even without that human) the window assemblage controls what effects the opening has on the interior of the building: how much light is let in, the amount of air and noise that is admitted or released, levels of safety and security. In short, windows modulate atmosphere and comfort. Although there is a vast technical literature on the relationship between windows and indoor thermal comfort (and also security), this has not featured in cultural histories of the window. An exception is Sandy Isenstadt's account of the American suburban window. He shows that part of the history of the suburban plate glass window is the contentious relationship between its visual and atmospheric effects. Isenstadt shows that during the late nineteenth- and early twentieth-century in both Britain and America there was considerable discussion among architects, designers and home-owners about the relationship between windows as a mechanism for delivering a view into a home and the effects of expanses of glass on the look and feel of domestic spaces. Specifically, he presents evidence of the ways in which a move to larger windows was seen to erode a sense of warmth in the home. Interestingly, the examples that Isenstadt gives are not about the impact of large plate glass windows on actual temperatures and air circulation (atmosphere in narrow sense), but more on the 'sense' of warmth and enclosure (atmosphere in the wider sense). For instance, Isenstadt cites a quaint example derived from an 1864 issue of Harper's Weekly. A Mr. Rogers was reported to have caught a cold simply from dining near a plate glass window. It was not a draft or chill from the (apparently closed) window that produced the cold-inducing atmosphere, but simply Mr Roger's 'force of imagination'.

One cannot separate off the ability of the window to attach itself evermore emphatically and narrowly to the experience of the 'view', from the elaboration of other non-window technologies given over to the management of atmosphere (temperature and circulation). For example, if we return to Le Corbusier's distended eye atop his flayed high-rise apartment (see Figure 8.2), we see that it is supported by a range of other technologies, one of which is called 'exact air' (a mechanical delivery system of ventilation, heating and cooling). The emergence of the modernist window as a viewing machine occurred in unison with the development of other technologies intent on creating the home as a 'climatic fortress' (Shove 2003). Indeed, it may well be that the rise of the window as an expansive viewing technology was dependent upon transferring its role as an aperture of atmosphere to other technologies of comfort.

While the windows of Red Road cannot but help be viewing devices, they are also called upon to play an important part in the control of atmosphere, for there is no 'exact air' system in Red Road. The management of interior temperature and ventilation happens by way of orchestrated activities involving the in-built heating system, ad hoc (resident supplied) technologies of heating and cooling, and the windows, which as we have noted can open a shut with a tilt-and-turn system. How to manage atmospheric comfort has been a concern to both housing managers and residents alike. This is nowhere more evident than in the quest to keep warm. In the history of the estate, three different domestic heating systems have been provided by the local housing authority. These systems have been subsidised by residents themselves, with technology ranging from portable electric and oil heaters, to kerosene burners, and including the construction of elaborate fake hearths.

As an opening, a window does its essential work in relation to that which is not window, the wall. In the case of the Red Road this relationship has shaped much, not only about how its windows can perform, but the entire building. This was given explicit expression by one Red Road resident whose first (unsolicited) mention of the windows of Red Road was inextricably linked to talking about another Red Road material, that of asbestos:

> We had asbestos, ay. See, that's all asbestos up there. Up there, next to the blinds. That ceiling. They've not took it away yet. They say because ... you see we always wanted [new] windaes, see.

The cladded steel-frame structure of Red Road was, for its time, considered novel by UK standards, given that most tower block construction was steel reinforced concrete (see Horsey 1982, Jacobs, Cairns and Strebel 2007). It was this novel building technology that resulted in the introduction of asbestos, which was used as the fire insulating material. As a result of the presence of asbestos, repair and replacement work of various kinds has produced specific safety issues for professionals and residents alike. It has above all meant that the windows in Red Road cannot economically or safely be replaced and so remain the original single glazed steel frame windows. Residents are forced to live with an ageing and worn window technology.

The inability to up-grade Red Road's windows has, in the light of recent Scottish Government policy intensified the 'problem' of Red Road as a satisfactory housing solution. In 2002 the Scottish Executive issued its Sustainable Scotland statement, which set down specific goals for windows including the requirement that they be double-glazed with low emissivity glass. Framing for windows should be uPVC, timber or aluminium, and that all fittings allow for ease of operation and safety. Shortly after, the Scottish Executive published its Scottish Housing Quality Standard, which further expanded the regulatory framework for windows. The Standard required that all properties be free from serious disrepair, energy efficient, and healthy, safe and secure. All social landlords were required to produce housing quality delivery plans that outlined a programme of action ensuring the standards were met by 2015. Attached as it is to this body of regulation, the Red Road window is increasingly experienced as a 'failure' not only by those close to hand (the residents) but also by more distant government agencies who are obliged to comply to these standards (Latour (writing as Johnson), 1988). Put another way, this single-glazed, steel-frame, asbestos-encased window 'fails' in and of itself, and because it is out of alignment with a re-specified national standard for window performance.

Conclusion

In recent moves to materialise vision, scholars have called upon Roland Barthes interpretation of the photograph as a laminated object: both material and visual. He uses the example of the window and its view to illustrate his point: like 'the windowpane and the landscape' he says (Barthes 1993: 6). We hope that this chapter has demonstrated how this is an apt metaphor, and in ways that go beyond the more strictly semiotic point that Barthes was originally making. This chapter has looked into the relationship between visualities and materialites in the context of modernist residential high-rise housing, its windows and its other practical orders of seeing. The story we have constructed is part cultural history, part visual studies, part sociologies of technology in action. Our close observations of Red Road's windows and its other practices of visuality are not, like many of the social sciences to precede it, bent upon simply inserting users (and their opinions) into the frame of buildings: soliciting resident thoughts and feelings of high-rise housing, its windows, and its surveillance technologies. Our intent has been in another direction. We have tried to capture something of the way in which building users and visual technologies work together to hold or not a building in place. In the case of residents we have focussed on how they and their windows work together to perform 'a high-rise view'. Such a view event was, as we demonstrated, central to the original Le Corbusian vision of high modern living. Yet it was imperfectly materialised in welfarist mass housing schemes. And, in the twilight days of those housing visions, when the very fabric of the housing built is decaying and breaking through disinvestment and age, performing a high-rise view is a thoroughly contingent and compromised event. Such difficulties arise not simply because the windows have failed, but residents too no longer domesticate the window technology as they once might have ('proper' curtains cannot be fitted, windows are left uncleaned, smoke stains). Similarly, when we turn our attention to the work of concierges in Red Road we come to see the ways in which a range of embodied and virtual visualisation practices work towards holding Red Road together as a safe, if somewhat dilapidated, housing complex. But, of course, Red Road is coming down and our final thought, must address this radical fact of de-materialisation. It is of note that much of the research on technology focuses on how new technologies undergo processes of 'integration' into everyday social norms (Shove and Hand 2007). What we have witnessed with the visualisation practices of Red Road may well have been the co-production of 'disintegration'.

Acknowledgements

This chapter is the result of research supported by the Arts and Humanities Research Council.

References

Barthes, R. (1993), *Camera Lucida: Reflections on Photography* (London: Vintage).
Benjamin, W. (1999), *The Arcades Project* (Cambridge: Belknap Press).

Colomina, B. (1994), *Privacy and Publicity: Modern Architecture as Mass Media* (Cambridge: The MIT Press).

Cosgrove, D. (2008), *Geography and Vision: Seeing, Imagining and Representing the World* (New York: I.B. Taurus).

Dunleavy, P. (1981), *The Politics of Mass Housing in Britain, 1945–1975* (Oxford: Clarendon Press).

Friedberg, A. (2006), *The Virtual Window: From Alberti to Microsoft* (Cambridge: The MIT Press).

Gieryn, T.F. (2006), 'City as Truth-spot: Laboratories and Field-sites in Urban Studies', *Social Studies of Science*, 36:1, 5–38.

Hand, M. and Shove, E. (2007), 'Condensing Practices: Ways of Living with a Freezer', *Journal of Consumer Culture*, 7:1, 79–104.

Harris, D. and Ruggles, D.F. (eds) (2007), *Sites Unseen: Landscape and Vision* (Pittsburgh: University of Pittsburgh Press).

Heath, C. and Luff, P. (2000), *Technology in Action* (Cambridge: Cambridge University Press).

Horsey, M. (1982), 'The Story of Red Road Flats', *Town and Country Planning*, July/August, 177–179.

Isenstadt, S. (2007), 'Four Views, Three of Them Through Glass', in *Sites Unseen: Landscape and Vision*, (eds) D. Harris and D.F. Ruggles (Pittsburgh: University of Pittsburgh Press), 213–240.

Jacobs, J.M. and Cairns, S. (2010), 'Ecologies of Dwelling: Maintaining High-rise Housing in Singapore', in *The New Blackwell Companion to the City*, (ed.) G. Bridge and S. Watson (New York: Wiley-Blackwell), 79–95.

Jacobs, J.M., Cairns, S. and Strebel, I. (2007), 'A Tall Storey ... but a Fact Just the Same: The Red Road High-rise as a Black Box', *Urban Studies*, 44:3, 609–629.

Jacobs, J.M., Cairns, S. and Strebel, I. (2008), 'Windows: Re-viewing Red Road', *Scottish Geographical Journal*, 124:2–3, 165–184.

Jephcott, P. (1971), *Homes in High Flats: Some of the Human Problems Involved in Multi-Storey Housing* (Edinburgh: Oliver & Boyd).

Kaplan, S. and Kaplan, R. (1978), *Humanscapes: Environments for People* (North Scituate: Duxbury Press).

Kelling, G.L. and Wilson, J.Q. (1982), 'Broken Windows: The Police and Neighborhood Safety', *Atlantic Magazine*, online archive <http://www.theatlantic.com/magazine/archive/1982/03/broken-windows/4465/>.

Kraftl, P. and Adey, P. (2008), 'Architecture/Affect/Inhabitation: Geographies of Being-in-Buildings', *Annals of the Association of American Geographers*, 98:1, 213–231.

Latour, B. (as Johnson, J.) (1988), 'Mixing Humans and Non-humans Together: The Sociology of a Door-closer', *Social Problems*, 35:3, 298–310.

Law, J. (2003), 'Ladbroke Grove, or How to Think about Failing Systems', (Lancaster, Lancaster University: Centre for Science Studies), <http://www.comp.lancs.ac.uk/sociology/papers/Law-Ladbroke-Grove-Failing-Systems.pdf>.

Lebas, E. (2007), 'Glasgow's Progress: The Films of Glasgow Corporation 1938–1978', *Film Studies*, 10:Spring, 34–53.

Le Corbusier 1967 (1933), *The Radiant City* (London: Faber and Faber).

Le Corbusier (1987), *The City of Tomorrow and its Planning* (New York: Dover Publications).

Macbeth, D. (1999), 'Glances, Trances, and their Relevance for a Visual Sociology' in *Media Studies: Ethnomethodological Approaches*, (ed.) P.L. Jalbert (Lanham: University Press of America & Int. Inst. for Ethnomethodology and Conversation Analysis), 135–170.

Markus, T.A. (1967), 'The Function of Windows – a Reappraisal', *Building Science*, 2:1, 97–121.

Markus, T.A. (1993), *Buildings & Power: Freedom and Control in the Origin of Modern Building Types* (London: Routledge).

Newman, O. (1972), *Defensible Space: Crime Prevention through Urban Design* (New York: The Macmillan Company).

Orlikowski, W.J. (2005), 'Material Works: Exploring the Situated Entanglement of Technological Performativity and Human Agency', *Scandinavian Journal of Information Systems*,17:1, 183–186.

Perrot, M., Ariés, P. and Duby, G. (eds) (1990), *A History of Private Life, Volume 4: From the Fires of Revolution to the Great War* (Cambridge: Belknap Press).

Pinder, D. (2005), *Visions of the City: Utopianism, Power, and Politics in Twentieth-century Urbanism* (Edinburgh: Edinburgh University Press).

Rainwater, L. (1972), *Behind Ghetto walls: Black Family Life in a Federal Slum* (Chicago: Aldine).

Reichlin, B. (1984), 'The Pros and Cons of the Horizontal Window', *Daidalo*, 13:2, 64–78.

Rose, G., Degen, M. and Basdas, B. (2010), 'More on "Big Things": Building Events and Feelings', *Transactions of the Institute of British Geographers NS*, 35:3, 334–349.

Shove, E. (2003), *Comfort, Cleanliness and Convenience: The Social Organization of Normality* (Oxford and New York: Berg).

Strebel, I. (2011), 'The Living Building: Concierges and the Vital Work of Block Checking Residential High-rises', *Social and Cultural Geography.*

Wood, D. and Beck, R.J. (1994), *Home Rules* (Baltimore: The Johns Hopkins University Press).

Yancey, W.L. (1978), 'Architecture, Interaction and Social Control' in *Humanscapes: Environments for People*, (eds) S. Kaplan and R. Kaplan, (North Scituate: Duxbury Press), 293–307.

Chapter 9

Melancholic Memorialisation:
The Ethical Demands of Grievable Lives

Karen Wells

Monuments may be interpreted as sites of sedimented history (Miles, 1997). Scholarly interest in the material culture of memorialisation has primarily focused on the role of monuments in the formation of national history. In this view, the nineteenth century era of monumentalisation is viewed as inseparable from the project of building the bourgeois nation and instantiating it in the material culture of the city (Huyssen, 2003; Therborn, 2002; Atkinson and Cosgrove, 1998; Osborne, 1998; Johnson; 1995). The place of war memorials in this project of nation building is more contested. While the majority of these memorials, certainly up until the First World War, commemorate battles and generals, the recognition of the deaths of soldiers and civilians have increasingly been put at the centre of monuments and memorials. After the 1914–18 war, war memorials valorised the deaths of the common people, for instance in the inauguration of the tomb of the unknown warrior, albeit always in ways that wrote their deaths into patriotic narratives. The debate over how to honour the victims of fascism in Germany at the end of the Second World War has most often been expressed in terms of the impossibility of representation (Young, 1993) but it is also a debate about how to break with a tradition of monumentalising that until that point had been unabashedly patriotic, nationalist and for the most part located with the 'great and the good'. Arguably another era of commemoration was inaugurated at the end of the Second World War, one which increasingly insists not only on the rights of the common people to be publically but also one in which the myth of a unified national identity unravels as the contested claims of vernacular histories struggle to make themselves visible in a public landscape of memorialisation (Dwyer and Alderman, 2008; Dwyer, 2000; Nora, 1998) in which 'no event since World War II has been fully assimilable to a unified national memory' (Nora, 1998: 616). These efforts were supplemented by activists and municipal socialists who sought to inscribe the urban landscape with significant names and dates in anti-racist and working-class struggles through the erection of monuments, the commissioning of murals, and the renaming of streets (Dwyer and Alderman, 2008; Wells, 2007a).

Notwithstanding the democratic impulse animating this new turn in public practices of memorialisation and the often contested interpretations of events that they may momentarily settle or at least make visible, they share a focus on events of national significance and often on 'great men' (Dwyer, 2000). In contrast, the memorialisations of those whose deaths do not in some way reverberate in national history are confined to acts of private grief in the quasi-public space of the cemetery. However, a relatively recent move towards commemorating traffic deaths and murder through the placement of flowers, ghost-bikes, and other artefacts has started to attract the attention of scholars of religion and material culture. This literature treats such commemorations as a kind of displacement of the expression of loss from culturally sanctioned sites, principally graveyards, to sites that are felt by the bereaved to be saturated with meaning. These new sites might include shrines (Maddrell, 2009) but are frequently located at the site where the person died. The literature on these new practices of grieving is largely concerned with the personal and spiritual implications for how people come to terms with loss (Petersson and Wingren, 2011; Clark and Franzmann, 2006). Some of this literature hints at the political significance of making death by 'road trauma' or murder visible in public space, for example by opening up the scripts of a destructive masculinity that are often implicated in both road trauma and murder (Hartig and Dunn, 1998). The majority of these kinds of memorials are temporary, and often mark the location of sudden and unexpected death, particularly of children and young people killed in traffic accidents.

This chapter is concerned with a third type of memorialisation, that sits somewhere between memorials that commemorate public events and memorials that mark personal loss onto the public landscape (Sidaway and Matell, 2007; Simpson and Corbridge, 2006). If the former, even when contested, form part of the narrative of public recognition of particular events and people and the latter are laden with poignancy precisely because they only matter to those for whom the fabric of everyday life has now unravelled, the kinds of memorials that I am concerned with in this chapter are of deaths that haunt the divide between the personal and the political. On the one hand, like those deaths commemorated by ghost bikes and cellophane wrapped flowers, these are deaths in a minor key. However, because the circumstances in which they died potentially signified political and economic inequalities, for a time at least, these personal tragedies entered political discourse. In this respect they are perhaps more directly comparable to the memorials to those who died in school shootings (Sci, 2009; Doss, 2002), or to the memorial to the victims of the bombing by a lone far-right American nationalist in Ohio in 1995 (Doss, 2002) or those who died in the attack on the World Trade Center on September 11 (Sidaway and Mayell, 2007). Those tragedies demand memorialisation because of the scale of the loss and the national demand for catharsis (Doss, 2002; Foote, 1997). In contrast the memorials I discuss in this chapter have to struggle to keep their losses in public view.

In her review of research on deathscapes, Lily Kong points to the need for 'closely documented detailed empirical observations' (1999: 8) of sacred and contested deathscapes. It is the aim of this chapter to describe empirically the processes of memorialisation and to elaborate from this description a schema for understanding the relationship between tragedy, memorialisation and politics.

The two deaths whose memorials are the subjects of this chapter ostensibly have little in common except for their spatial proximity. One took place two years later and less than a mile away from the other, and in very different circumstances. At first sight there seems little reason other than proximity to use these two memorials to think about the ethics and politics of responding to violence in the city. However, risk, violence and insecurity in the city are spatially distributed (Mesev, et al., 2009; Ruston, 2009). It is not coincidence that these two killings happened so close by one another. Secondly, although the handful of academic analyses of one of my examples – the shooting of Jean Charles de Menezes – locate this death in the context of the 'war on terror' and state security (O'Driscoll, 2008; Vaughan-Williams, 2007), it can also be considered as part of police violence towards subjects in working-class neighbourhoods;[1] it is simply inconceivable that he would have been shot in this fashion had he been followed onto the underground in a more wealthy part of the city.

Furthermore, which is the focus of this chapter, close attention to the visuality and materiality of these memorials as they changed over the period, from the initial shock of their deaths through to their final memorial forms, suggests that they share not only proximity but also practices of memorialisation that move from shock through melancholia to mourning and that in the liminal space of what might be called melancholic memorialisation loss is opened up to the world and makes political claims upon it.

My argument proceeds in the following way: I begin with a description of the circumstances in which the two people commemorated in these memorials died. I then describe how the initial response to these tragedies were materialised and visualised in artefacts and images. I suggest that the initial response to tragedy signifies unfocused despair and rage. I then discuss how a process of ordering and containment was gradually brought to bear on these materialising and visualising practices but that in this period, despite these processes of containment, the memorials that were constructed were still temporary and made from ephemeral materials (Doss, 2002). I propose that this period may be understood as *melancholic memorialisation*. In this period the interpretation of events and their significance for wider political questions of the unequal spatial and social distribution of violence are at the forefront of practices of commemoration (Simpson and Corbridge, 2006). This moment, more or less extended, of melancholic memorialisation gives way, in turn, to the construction of a permanent memorial that closes down the political space opened up first by tragedy and then by melancholia. This final testimonial to loss may be resignified by events, interventions and practices that return loss once again to melancholic memorialisation and reanimate its political potential.

---

1   Since 2000 126 people have been killed by the Metropolitan police including 12 shootings (<http://www.inquest.org.uk/>.

The Killing of Jean Charles de Menezes

On July 7th 2005 four bombs exploded in London killing 56 people and injuring over 700 others. The bombs were exploded by three teenagers and a 30-year-old. They were all killed by their own explosive devices (HOC, 2006; Tulloch, 2006). Two weeks later, there was another attempted explosion of four devices on the London Underground and a bus. The devices failed to detonate and the four terrorists[2] escaped unharmed.

The block of flats in Tulse Hill, south London, where one of the would-be July 21st bombers, Hussein Osman, was thought to live, was put under surveillance. Reportedly the officer who was watching the block left his position to urinate. When he returned he saw a man leaving the block who he thought might be Osman, it was in fact de Menezes, and he was then followed by two undercover police officers. At Stockwell tube station he was followed onto a train by two armed police officers who shot and killed him. In the immediate aftermath of his killing, the police made a number of claims, later to be found to be untrue and/ or contradicted by eye-witnesses to explain how they came to believe that de Menezes was Osman and why they believed that their life or the lives of others were in immediate danger from Hussein/Jean sufficient to justify killing him (IPCC, 2007a, 2007b; O'Driscoll, 2008; Vaughan-Williams, 2007).

A campaign to have the police officers charged with murder, the Justice4Jean campaign, began shortly afterwards. An inquest into his death held in December 2008 returned an open verdict after the jury was told by the coroner that they could not pass a verdict of unlawful killing. In November 2009, the de Menezes family accepted compensation from the Metropolitan police of £100,000 and ended their legal action against the police.

The Murder of Billy Cox

Billy Cox was 15 years old when he was killed by a still unidentified person at his home on the Fenwick Estate, Clapham North, South London, where he lived with his parents and 13-year-old sister. The police believe that he was shot by somebody he

---

2   I recognise that the word 'terrorist' is problematic. One of the bombers declared himself 'a soldier' in a war against the British state/people. In doing so he is following in a long tradition (not only, or even mostly, Islamic) of individuals who hope that political change can be produced through individual acts of violence against a population who are in the insurgents' view guilty of colluding with state-sponsored injustice. However, the use of 'terror' against people who become targets by the presumption that they support the incumbent government is always deeply problematic in its implicit (and nationalistic) contention that society is not already divided and conflictual. This chapter is not the place to discuss in depth the politics of terror but nonetheless, these terms do necessitate interrogation (see Jenkins, 2006; Gambetta, 2005; Pape, 2005).

knew in connection with drug-dealing. In a call for witnesses on the third anniversary of his death the police and Billy Cox's father told the press that they were confident that they know who committed the murder but that they have insufficient evidence to prosecute. An inquest into his death in 2008 returned a verdict of unlawful killing.

Figure 9.1   Memorial to Jean Charles de Menezes
Note: See also <http://www.guardian.co.uk/news/gallery/2007/oct/01/demenezes#/?pictur e=340527232&index=13>

Figure 9.2   Billy Cox's street name 'Remer' on his estate in the days after his death

The Memorials

From Despair and Rage ...

In the first days after the police shooting of de Menezes, some handwritten condolence notes, head and shoulder photographs and a poster were sellotaped to the wall next to the tube station entrance. This was the start of the accretion of notices, newspaper clippings, and photographs that was assembled into a monument attesting to the fact of his murder and insisted on the necessity of someone being held to account for his death. The memorial worked simultaneously as a site for grieving – fresh flowers were repeatedly replaced as they might be at a grave – and a site to witness injustice. The injustice of de Menezes death was located from the outset in a wider discourse of state injustice; for example, the press cuttings and statements linked this shooting to the deaths of people in police custody and to the war in Iraq.

The first stage of this memorialising of Jean Charles de Menezes was contained in a space next to the underground entrance. Despite its physical containment it almost fails as a unified sign, appearing to be chaotic and dispersed in its focus. It is perhaps because it was so much easier to locate his killing within the sphere of the political that the raw expression of focused loss that, as we shall see, is evident in the first days of memorialising after Billy Cox's death, was less evident at Stockwell. In the days after Billy Cox's murder the slogan 'RIP Remer' (his 'street' name) was spray-painted or chalked all over the estate in the same large, hastily drawn letters (Figure 9.2). A stencil of his portrait was spray-painted on the ground and walls of the estate repeatedly (Figure 9.3). The effect was very haunting and seemed to be intended to establish the whole estate as a site of grief. In amongst the many sentimental messages like 'heaven needed an ange' so they took you' were the words 'Da press chat shit' and 'shame that you had to be dead to be famous', perhaps hinting at something other than a response to his murder as a personal tragedy. Perhaps these comments also spoke to and contested wider political discourses about feral boys and feckless parents and the impossibility of 'being somebody' in the context of poverty; and perhaps to a situation where to 'be somebody' has been increasingly aligned with celebrity and 'gangsta' culture (Hatttam and Smith, 200; Adams, 1999).

... to Melancholia

These inchoate productions that did not quite amount to a sustained commentary on what has happened, soon gave way to a more orchestrated assembling of texts, artefacts and images. The slogans that had been scrawled all over the estate and the stencils of his face were quickly painted over, presumably by the local council anti-graffiti team, and the site of memorialisation became focused on the wall adjacent to his home. An armchair was put in front of the wall and covered in graffiti. Candles and flowers were placed around it. It resembled many such memorials

Figure 9.3   A stencil painting of Billy Cox's portrait on the walkways of his estate

Figure 9.4   The final memorial to Billy Cox

that appear at the sites of murders and traffic accidents, more elaborate than most but similar in the tone of the comments. A few texts, the use of his street name and a reference to him as a 'fallen soulja' pointed to his reported gang affiliation with the 'Clap Town Kids'. This is the moment that I am suggesting might be captured by 'melancholia' and it is this moment, analogous to the moment between a burial and the erection of some permanent marker on the burial site, when the meaning of a tragedy is located on the border between private grief and public justice. In other words it is at this moment when tragedy is open to the possibility of demanding a political response (Simpson and Corbridge, 2006). It is this moment when a demand for explanation, retribution, or recompense is made.

The struggle between those who want to keep the presence of the dead in public view and the authorities who want to hide or contain the loss is evident in both memorials. London Transport (the body responsible for underground stations) tried to get the memorial dismantled, citing the need to redevelop the station. They suggested that a plaque to his memory should be sited inside the station. The family resisted this move, claiming that it would take the case out of public view and make it less likely that justice would be done. In the event the memorial became more contained and organised as it began the transition from a site of despair through melancholia to a place of mourning. By July 2006 the shrine resembled the newspaper kiosk that it stood next to. The construction of a banner above the shrine and two thin 'walls' to either side extended the space that the shrine took up but at the same time, by making it a more permanent looking structure, also made it appear to recede into the surrounding landscape. The messages continued to link his death to other injustices, both deaths in police custody and the war in Iraq.

Similarly, the local authority has repeatedly threatened to erase the permanent memorial to Billy Cox. At the time of writing they have claimed that they carried out a petition of the residents of the estate in the weeks after the mural was first painted, and had discovered that a majority of the residents want the mural relocated. Their intention is now to commission a replica of the mural to be painted onto boards that will then be fixed onto the football pitch which is at the centre of the estate. The original painting will then be erased.

Mourning

On the day that Billy Cox was buried a mural was painted on a wall adjacent to his house.[3] The armchair was taken away and the slogans that had been scrawled on the walls around it were painted over then inscribed again, this time in more carefully drawn letters. The aesthetic of the mural is sacred rather than political; it seems to gesture towards a religious iconography (Maddrell, 2009) enhanced by its luminosity at night as the headlights of cars illuminate it briefly before taking the

---

3   The mural was painted by Dane, a professional graffiti artist with VOP (<http://www.vopstars.com>) and funded by money raised by Billy Cox's friends from a door-to-door collection on the estate.

sharp left that it stands in front of. At the foot of the mural there are fresh flowers and candles, the latter always alight. This seems to be intended to be the final step from tragedy to mourning. The struggle to have his death inserted into political discourses was quickly lost and the space for melancholic memorialisation was very brief. This is not a site that materially or visually readily signifies rage.

The transition from melancholia to mourning – the recognition of the failure to make somebody take responsibility for his death – is materialised in the case of Jean Charles de Menezes by the erection of a permanent memorial to him outside of Stockwell station.[4] This memorial was unveiled on January 7 2010. It is a mosaic with an image of de Menezes portrait at the centre. Above this image are the words 'rest in peace' and below it 'INNOCENT/Jean Charles de Menezes/ Born Gonzoga MG Brazil 07.01.1978/Shot dead here 22.07.2005. Sadly missed'.

Facing Death

From the moment that the process of commemoration began the faces of the killed subjects were central to the practices of memorialisation in both instances. The wall by Stockwell tube station which developed into the temporary and then the permanent site of the memorialisation of Jean Charles de Menezes death began with an image of his face at its centre. Likewise, as I have already mentioned, the ground and walls of Billy Cox's estate were covered, within days of his death, by the repeated stencil painting of his face. The final memorial to him is a seven-foot tall head and shoulders portrait. The face of Jean Charles de Menezes was beamed onto the walls of the Houses of Parliament two years after he was killed in an effort to keep the struggle for someone to be brought to account for his death in the public eye.

In *Precarious Life*, Butler explores Levinas's claim that the 'face', makes 'moral claims upon us, addresses moral demands to us, ones that we don't not ask for, ones that we are not free to refuse' (2004: 131). Butler's interest is in representation of loss in public space in the context of the US 'war on terror' and its aggression against Afghanistan and, by extension, Islam. Although the connections between Jean Charles de Menezes' death and the 'war on terror' (see Vaughan-Williams, 2007) are simultaneously clear (because he was killed instead of another man who was subsequently jailed for 40 years for his part in the July 21 bombing plot) and faint (because he was not that man), it is not so much that connection that I am concerned with here as with the deployment of the face as a call on the spectator to recognise her/his responsibility for the life (and therefore the death) of the other. This deployment of the face is well established in practices of memorialisation, cross-culturally. Rozzane Varzi (2006: 24) in her ethnography of youth and martyrdom in Iran, describes the haunting quality of the photographs

---

4   The mosaic was made by a local artist, Mary Edwards, with the support of the de Menezes' family.

of young men on the graves at the cemetery of the war martyrs, Isfahan. She speaks of wanting to be 'protected' form their stares. Similarly, Levinas speaks of 'those eyes, which are absolutely without protection, the most naked part of the human body' (1990: 8). There is something uncanny about looking at reproductions of the face of the dead. Barthes's (2000) observation that the photograph always has the mark of death on it is heightened when the image is someone now dead because the image in some way holds the empty promise of resurrection (Rose, 2007: 46). The images of young men and boys violently killed in the city that are re-circulated in news stories each time another young person is violently killed has this same sense of the simultaneous promise and failure of the dead to speak. This promise/failure of the photograph compels the spectator to account for the life and death of the dead person; to explain it and signify it. Davis in his reading of Levinas contends that death 'is an interruption to the production of meaning' (Davis, 2004: 77) and that 'the other is a producer of signs to which death puts an abrupt end' (2004: 78) but at the same time that 'Because I am invested with responsibility for the other, the death of the other is necessarily my affair' (2004: 81).

This deployment of the face then, in practices of memorialisation, may be thought of as an attempt to produce signs on behalf of the dead that make an ethical claim on the spectator; one that compels the spectator to ask at the very least, 'what happened that caused this death' (Sontag, 2003). The inscription of the face of the dead in public space is insisting in an explicit (in the case of political campaigns like de Menezes) or implicit fashion that their death is a matter of public concern. In doing so these images interpellates passers-by as a community of spectators/witnesses. To refuse this demand is arguably unethical (Rose, 2007: 47). Although, as with other images of tragic death, as Rose remarks, 'We don't know what to do once we've seen them, apart from feel something' (Rose, 2007: 50).

The Politics of Melancholia

Drawing on Freud's essay 'Mourning and melancholia' several poststructuralist theorists have explored the political space produced by melancholia (Mookherjee, 2007; Gilroy, 2004; Butler, 2004; Eng and Kazanjion, 2003). Melancholia describes a response to loss in which the griever cannot incorporate the loss of the lost object/person. In mourning, the repetition of remembering the lost object/person enables the incorporation (in memory/consciousness) and through this the letting go of the lost person. In melancholia there is some obstacle to this process of incorporation and giving up and the melancholic continues to revisit the site of loss in unproductive ways (Freud, 1917). However, what maybe, in Freud's terms, unproductive for the individual psyche of the traumatised may be, because of its refusal to foreclose the political signification of loss, productive in making political claims from apparently personal tragedy (Eng and Kazanjian, 2003: 3).

If melancholia is a refusal to let go of the lost object, then this moment of memorialisation in which the commemoration of loss is fragmented, assembled, incomplete, made from found materials may be thought of as a heightened moment when the personal crosses the border into the political. In similar terms to this idea of melancholic memorialisation, drawing on Veena Das', work in India, Jill Bennett conceives of grieving as:

> the process of inhabitation in the wake of loss, a process of coming to terms, which does not imply a decathecting of desire from the lost object so much as a reenvisioning of the world ... insisting on a recognition of the material encounters that occur in the wake of loss, and focusing not simply upon the private aspects of trauma but on the transformation of space itself. Grief is thus thought of in terms of an outfolding into the world that must be remade in the aftermath of tragedy. (Bennett, 2005: 68).

What I take this 'outfolding into the world' to be is precisely a demand that violent death is not absorbed into personal grief but is expressed in public and articulated in the language of politics – power and justice.

Innocent

The period of melancholic memorialisation does not, of course, happen without contestation. It is necessitated precisely by the fact that there is dissonance between how (at least) two communities signify these deaths. One of the expressions of this contestation is in whether the victim deserves to be honoured, or even, possibly whether they were in some way culpable in their own death. The refusal to move on, and instead to maintain the presence of the dead in public space, is more easily accepted as a legitimate (if still contested) demand when the dead person can in no way be deemed culpable in his/her own death. We might therefore compare the media and police response to these two deaths. In both cases the question of whether they were innocent – not so much in the sense of having provoked another person to kill them, but rather in the sense of being legitimate (literally, legal in Jean Charles de Menezes' case), was an important factor in establishing the possibility that their loss in some sense demanded a political response. In the hours and days after Jean Charles de Menezes's killing, accounts by the police were circulated through the media with the intention of making him seem in some way not a legitimate person. Three claims were made: one was that his 'indefinite leave to remain' stamp in his passport was a forgery and that he was therefore 'illegal'; a second was essentially that he acted like a guilty person (wearing a bulky jacket to conceal his body, jumping the barrier when challenged at the tube station, and running onto the train). Frequent recourse to the unassailability of the testimony of visual witnesses – both electronic and human – were made in the hours after his death. This unquestioning of eye-witness testimony is increasingly becoming part of news media (Chouliaraki, 2010; Rose, 2007: 55). It transpired that he was wearing a light denim jacket, that he used his travel pass to go through the ticket barrier and that the train he ran for was about to leave the station. In any case, the point is that the police reports were intended to erode his credibility and in some way thereby make his death less political. His family and the *Justice4Jean* campaign were engaged in similar discursive strategies intended to restore his credibility, for example, insisting that he had the right to be in the UK either because the Home Office were wrong when they said that his ILR stamp was not one they issued or that he last entered the UK via the Republic of Ireland and therefore was legal at the time of his death; that he did not vault the barrier, etc. They also repeatedly spoke of him as hard-working, intelligent, family orientated, and kind. Neither side insisted that whether Jean Charles de Menezes was of good character or not, whether his status was undocumented or not, is irrelevant to the determination of whether the tragedy of his death demanded a political response. His family were probably right in thinking that in order to make a political demand they first had to establish his credibility. Having established his credibility, his respectability, they could then use the melancholic space that his innocent death opened up to argue that justice should be done. It is instructive that the final memorial to him at Stockwell station bears the legend: innocent.

Billy Cox's friends who tagged his estate after his murder were also engaged in discursively constructing Billy Cox as a good person, and their claims were also contested. Some of this discursive work was done on the estate and the walls near his home that eventually became the site of the permanent memorial. Statements like 'god needed an angel so they took you', 'only the good die young', 'another fallen soulja', and 'never forgotten' insisted on the grievability of his death. The media, relaying this witnessing, added its own commentary that situated his murder in a familiar tale of poverty, youth gangs and so-called 'black' gun crime. Here again, the discourse of two apparently opposing sides shares some common ground: to be grievable is to be innocent. Although the death of young men and boys had become a recurrent topic in news media by the time Billy Cox was shot, the terms of this discourse are very narrow. For the most part violence is explained, tautologically, as an effect of gang affiliation. In the context of this discourse the victims of gang-related crime, if they are themselves gang affiliated (or presumed to be) bear the stain of guilt. However, the impossibility in this context of establishing Billy Cox as respectable and therefore innocent means that it is also impossible to establish the melancholic space which insists that tragedy is political, or that it outfolds to the world. In this context the public mourning for Billy Cox quickly resolved into the production of a mural that signified his death as more sacred than political.

The Politics of Justice

What might be meant by politics in these contexts? In both these instances, the demand for explanation and retribution were not framed by politics but by law (to the extent that these two discourses can be separated from one another); justice would be seen to be done if the killers were brought to account in law, charged with murder. In neither case did this happen. Since Jean Charles de Menezes was killed by the police who attempted to legitimate their error with reference to the demands of security the possibility of inserting his death into a political space, of making a political community from the demand that someone be held accountable for his death, was amplified. Despite this possibility, and notwithstanding some initial attempts to associate his killing with other deaths at the hands of the police, the political space that developed around his death never expanded beyond legal justice and even this limited demand was never met; as with many other deaths at the hands of the police, the police succeeded, with the active support of the rest of the governing apparatus, in insulating themselves from the politics of justice from below. Indeed Cressida Dick, who supervised the surveillance that led to his death, was awarded the Queen's Police Medal in 2010. A financial settlement was made and his death returned to personal loss: sadly missed.

Conclusion

In this chapter I have shown that these two memorials – honouring two very different lives and their very different endings – share a set of practices. The initial inscription of loss was angry and bare. In the case of Billy Cox's death the tags all over the estate were like an explosion of words and images. The impression they made was of a scrawl repeated innumerable times as the writers ran through the estate determined to witness their friend's death. For Jean Charles de Menezes the start of his commemoration was also angry and bare but also very contained. Contained in the one case and dispersed in the other, but sharing an aesthetic of rawness, of urgency, of making do. This initial impulse to do something, to quickly give expression to a raw anger, was later organised into a more coherent but still provisional assemblage of texts, objects and images. The location for this assemblage of objects begins to resemble a shrine; occupying the boundary between the sacred and the political. For Billy Cox the possibilities for making his death political (from below) were always slight, and the time between first loss and final memorial, that melancholic space that I am arguing is a productive political space, was brief. In both cases, the visual artistic aesthetic of the last memorial seems to stabilise and limit what is being asked for of the spectator. Despite the continuing ethical demand of the image of the face, what is demanded is now muted and slippery; perhaps only the recognition of a family's tragedy and the acknowledgement that somebody care(d/s) for them.

A political response to violent death is opened up by practices of melancholic memorialisation but then foreclosed, I am arguing, by the production of a highly stylised image that references artistic and religious aesthetic conventions and is made of permanent materials. The ethical demands made by the inscription of the subjects' faces on their final memorials are circumscribed by the earlier failure of the melancholic moment to tie these deaths to political discourse. The final version of the memorialising practices deployed to grieve these loses have interrupted the circuits of witnessing and testimony that began with the rough designs in the first days after their deaths. However whilst, in one sense, these memorials have laid matters to rest, the very continuation of their inscription in public space means that the matter does not quite end there. These are not after all graveyards. Fenwick estate is home to several hundred families and Stockwell tube is used by thousands of commuters daily.

It has often been remarked in one way or another that monuments are not seen unless some event causes their re-signification. Forgotten but not gone. In time these memorials too may lose their impact and fade into the urban landscape. However, perhaps the reason that monuments fade from view, except in the tourist landscape, is that what they are intended to signify is so obvious that it becomes literally un-remarkable. In contrast, the two memorials discussed here have no such straightforward signification and the difficulty of locating precisely what it is that these memorials signify also contributes to their continuing semiotic presence.

These memorials, for all their foreclosure of political claims, may still be taken as simply a statement of presence, a refusal of erasure. In this respect they undermine Butler's contention that for those whose lives 'on the level of discourse, … are not considered lives at all' (and I take that to include the lives of the poor), 'there will be no public act of grieving' (Butler, 2004: 36). Her argument, whilst valid at the level of official discourses of grievability, ignores the everyday practices of recognition that are enacted to publically mourn violent death in the working-class city. These practices, including the memorials discussed here, might be thought of as de Certeau's tactics (1984) or Lefebvre's representational spaces (1991, see also Wells, 2007b) In her commentary on Krzysztof Wodiczko's video projection on the murder of youths in teen violence in Charlestown, Sarah Purcell (2003) comments that the voices of the boys' mothers projected onto the bunker hill monument, are a vocal insistence that 'we are here' and we will be heard.

That monument, originally erected to honour the dead of the Battle of Bunker Hill, one of the sites of the North American War of Independence (1775–1783), was resignified through an artistic intervention that made the memorial an object of melancholic loss. For who is the 'we' that is 'still here'; not the dead boys whose mothers now bear witness to their deaths. The installation/monument is a refusal to let go of their dead sons. The mourning memorial, then, can be resignified, sliding backwards and forwards between a past that is 'declared resolved' and a past 'that remains steadfastly alive in the present' (Eng and Kazanjian, 2003: 4). The return of a memorial to a melancholic object may occur when its stillness is disrupted by some event: a new attempt to erase it; an anniversary, another death, an intervention. My contention then is that the analysis of memorials can unsettle their finality and restore them to melancholia so that they may continue to provoke us to ask political questions about the unequal distribution of violence, risk and (in) security in the contemporary city.

References

Adams, N.G. (1999), 'Fighting to be Somebody: Resisting Erasure and the Discursive Practices of Female Adolescent', *Educational Studies*, 30:2, 115.

Atkinson, D. and Cosgrove, D. (1998), 'Urban Rhetoric and Embodied Identities: City, Nation, and Empire at the Vittorio', *Annals of the Association of American Geographers*, 88:1, 28.

Barthes, R. (2000), *Camera Lucida: Reflections on Photography* (London: Vintage).

Bennett, J. (2005), *Empathic Vision: Affect, Trauma, and Contemporary Art* (Stanford: Stanford University Press).

Butler, J. (2004), *Precarious Life: The Powers of Mourning and Violence* (London: Verso).

Chouliaraki, L. (2010), 'Ordinary Witnessing in Post-television News: Towards a New Moral Imagination', *Critical Discourse Studies*, 7:4, 305–319.

Clark, J. and Franzmann, M. (2006), 'Authority from Grief, Presence and Place in the Making of Roadside Memorials', *Death Studies*, 30:6, 579–599.

Davis, C. (2004), 'Can the Dead Speak to Us? De Man, Levinas and Agamben', *Culture, Theory & Critique*, 45:1, 77–89.

de Certeau, M.D. (1984), *The Practice of Everyday Life* (University of California Press: Berkeley).

Doss, E. (2002), 'Death, Art and Memory in the Public Sphere: The Visual and Material Culture of Grief in Contemporary America', *Mortality*, 7:1, 63–82.

Dwyer, O.J. (2000), 'Interpreting the Civil Rights Movement: Place, Memory, and Conflict', *Professional Geographer*, 52:4, 660.

Dwyer, O.J. (2008), *Civil Rights Memorials and the Geography of Memory*. 1st ed. D.H. Alderman (ed.) (Chicago: Center for American Places at Columbia College Chicago).

Eng, D.L. and Kazanjian, D. (eds) (2003), *Loss: The Politics of Mourning* (Berkeley: University of California Press).

Foote, K.E. (1997), *Shadowed Ground: America's Landscape of Violence and Tragedy* (Austin: University of Texas Press).

Freud, S. (1957) [1917], 'On Mourning and Melancholia'. In (ed.) S. Freud *The Standard Edition of the Complete Psychological Works of Sigmund Freud: On the History of the Psycho-analytic Movement; Papers on Metapsychology, and other Works*. J. Strachey (trans.), (London: The Hogarth Press and the Institute of Psycho-analysis).

Gilroy, P. (2004), *After Empire: Melancholia or Convivial Culture?* (London: Routledge).

Hartig, K.V. and Dunn, K.M. (1998), 'Roadside Memorials: Interpreting New Deathscapes in Newcastle, New South Wales', *Australian Geographical Studies*, 36:1, 5.

Hattam, R. and Smyth, J. (2003), '"Not Everyone has a Perfect Life": Becoming Somebody without School', *Pedagogy, Culture & Society*, 11:3, 379–398.

House of Commons (2006), Report of the official account of the bombings in London on 7th July 2005 (London: The Stationery Office).

Huyssen, A. (2003), *Present Pasts: Urban Palimpsests and the Politics of Memory* (Stanford: Stanford University Press).

IPCC (2007a), Stockwell One: Investigation into the shooting of Jean Charles de Menezes at Stockwell underground station on 22 July 2005.

IPCC (2007b), Stockwell Two: An investigation into complaints about the Metropolitan Police Service's handling of public statements following the shooting of Jean Charles de Menezes on 22 July 2005.

Johnson, N. (1995), 'Cast in Stone: Monuments, Geography, and Nationalism', *Environment & Planning D: Society & Space*, 13:1, 51.

Kong, L. (1999), 'Cemetaries and Columbaria, Memorials and Mausoleums: Narrative and Interpretation in the Study of Deathscapes in Geography', *Australian Geographical Studies*, 37:1, 1.

Lefebvre, H. (1991), *The Production of Space* (Oxford: Blackwell).

Levinas, E. (1990), *Difficult Freedom: Essays on Judaism* (Baltimore: The John Hopkins University Press).

Maddrell, A. (2009), 'A Place for Grief and Belief: The Witness Cairn, Isle of Whithorn, Galloway, Scotland', *Social & Cultural Geography*, 10:6, 675–693.

Mesev, V., Shirlow, P. and Downs, J. (2009), 'The Geography of Conflict and Death in Belfast, Northern Ireland', *Annals of the Association of American Geographers*, 99:5, 893–903.

Miles, M. (1997), *Art, Space and the City: Public Art and Urban Futures* (London and New York: Routledge).

Mookherjee, N. (2007), 'The "Dead and their Double Duties": Mourning, Melancholia and the Martyred Intellectual Memorials in Bangladesh'. Special Issue the Material and Visual Culture of Cities, *Space and Culture*, 10.2 (May 2007): 271–291, 2006.

Nora, P. and Kritzman, L.D. (1996), (eds) 'The Era of Commemoration'. In *Realms of Memory: Rethinking the French Past* (New York: Columbia University Press).

O'Driscoll, C. (2008), 'Fear and Trust: The Shooting of Jean Charles de Menezes and the War on Terror', *Millennium (03058298)*, 36:2, 339–360.

Osborne, B.S. (1998), 'Constructing Landscapes of Power: The George Etienne Cartier Monument, Montreal', *Journal of Historical Geography*, 24:4, 430.

Petersson, A. and Wingren, C. (2011), 'Designing a Memorial Place: Continuing Care, Passage Landscapes and Future Memories', *Mortality*, 16:1, 54–69.

Purcell, S.J. (2003), 'Commemoration, Public Art, and the Changing Meaning of the Bunker Hill Monument', *Public Historian*, 25:2, 55.

Rose, G. (2007), 'Spectacle and Spectres: London 7 July 2005', *New Formations*, 62:1, 45–59.

Ruston, A. (2009), 'Isolation: A Threat and Means of Spatial Control. Living with Risk in a Deprived Neighbourhood', *Health, Risk & Society*, 11:3, 257–268.

Sci, S.A. (2009), '(Re)thinking the Memorial as a Place of Aesthetic Negotiation', *Culture, Theory & Critique*, 50:1, 41–57.

Sidaway, J.D. and Mayell, P. (2007), 'Monumental Geographies: Re-situating the State', *Cultural Geographies*, 14:1, 148–155.

Simpson, E. and Corbridge, S. (2006), 'The Geography of Things That May Become Memories: The 2001 Earthquake in Kachchh-Gujarat and the Politics of Rehabilitation in the Prememorial Era', *Annals of the Association of American Geographers*, 96:3, 566–585.

Sontag, S. (2003), *Regarding the Pain of Others* (New York: Farrar Straus and Giroux).

Therborn, G. (2002), 'Monumental Europe: The National Years. On the Iconography of European Capital Cities', *Housing, Theory & Society*, 19:1, 26–47.

Tulloch, J. (2006), *One Day in July: Experiencing 7/7* (London: Little Brown).

Varzi, R. (2006), *Warring Souls: Youth, Media, and Martyrdom in Post-revolution Iran* (Durham: Duke University Press). Available at <http://www.loc.gov/catdir/toc/ecip064/2005034368.html>.

Vaughan-Williams, N. (2007), 'The Shooting of Jean Charles de Menezes: New Border Politics?', *Alternatives: Global, Local, Political*, 32:2, 177–195.

Wells, K. (2007a), 'Symbolic Capital and Material Inequalities: Memorializing Class and "Race" in the Multicultural City', *Space and Culture*, 10:2, 195–206.

Wells, K. (2007b), 'The Material and Visual Cultures of Cities', *Space and Culture*, 10:2, 136–144.

Young, J.E. (1993), *The Texture of Memory: Holocaust Memorials and Meaning in Europe, Israel, and America* (New Haven: Yale University Press).

Chapter 10

Indifferent Looks: Visual Inattention and the Composition of Strangers

Paul Frosh

The majority of visual images circulating in contemporary media-saturated societies are experienced, it is probably safe to say, as fleeting and unremarkable ephemera. Vast numbers are encountered daily as objects of routine inattention and distraction. Barely registered consciously, our engagements with these images are 'unstoryable' in Paddy Scannell's sense (1996): so mundane that they are unworthy of notice or particular comment, and are hard to recall with any specificity.

Worryingly for those who create these images, our everyday disregard of them bears no obvious or direct correlation to the quantity of professional care and resources that have been lavished on their production. Research on media production – from news journalism to commercial advertising images – uncovers the commonplace fact that the images, texts and narratives which we habitually treat inattentively as a kind of unremarkable wallpaper are nevertheless the products of highly concentrated institutionalized forms of attention, routinely practiced by journalists, photographers, designers, editors and other cultural professionals (Dayan 2009; Frosh 2003).

Indeed, there appears to be something morally unseemly, or at least profoundly unsettling, in this lack of correlation. It may seem a blessing to be able to ignore the mass of unwanted appeals from 'intrusive' images, particularly advertisements and marketing messages. But what of images of others' suffering? Surely our moral connection to unfamiliar others depends upon a willingness to engage attentively the particular depictions of their personal distress? Doesn't the visual elicitation of moral responsiveness to distant sufferers, including by means of the kinds of 'impartial' spectatorship proposed by Adam Smith and elaborated more recently by Boltanski (1999), require forms of imaginative engagement – and concomitant modes of aesthetic representation – that are inimical to the uncommitted and indifferent routines of a culture of distraction (Moeller 1999; Chouliaraki 2006; Slovic 2007)?

The Attentive Fallacy

To be interested in our relations – particularly our moral relations – to images in conditions of inattentiveness invokes an obvious methodological paradox. How should one pay attention to inattention? Analysis requires specifying and focusing on images as distinct objects of interest and inquiry. But in so doing it threatens to treat them in a way that makes it difficult to account for our indifference towards them, as well as for their potential and actual indifference towards us. Except perhaps in terms of a 'bad object' of inquiry, in which conceptual and phenomenological slipperiness appear to compound the charge of ethical dubiousness.

Such a methodological paradox sits uneasily with the main assumptions of most of the disciplines that traditionally take visual media as their central concern. For instance, in the case of art history – long the paradigmatic intellectual context for the analysis of images – Michael Ann Holly has argued that there is 'a productive correspondence of rhetorical ideologies between image and text. Representational practices encoded in works of art continue to be encoded within their commentaries' (1990: 385). Offered as a critique of scholarly postures of objectivity, this claim nevertheless operates within the framework of a discipline where the identity and boundaries of the discrete individual work seem self-evident and straightforward, and where the corresponding issues of spectatorial and analytical attention – as that which the work unproblematically demands prior to its particular 'commentary' – do not pose any difficulties and go largely unnoticed. The idea that attention – curatorial, scholarly and intellectual (not to mention economic) – may be foundational to the special status enjoyed by art objects, art history and art criticism is simultaneously underscored and concealed by the deceptively commonsensical observation that, to quote Holly, 'paintings are, after all, meant to be looked at' (Holly 1990: 373). This assumption of an *intentionality* of attention – that something is looked at because it is intended to be looked at – allows Holly to develop a critique of how spectatorship implicates both scholarly and non-scholarly observers in the rhetorical tropes of the works they view, without at the same time moving on to a critique of the framing of art itself as a privileged regime of visual attentiveness.

What, however, of those objects, images and practices whose intentionality as objects of the attentive gaze is in doubt, or – even if they are meant to be looked at – are more often than not overlooked? How is one to seize this inattentiveness and to transform it into something stable enough for us to comprehend? How is one to capture and arrest the dynamism and mobility of that which eludes or resists our focused concentration, 'the rapid crowding of changing images, the sharp discontinuity in the grasp of a single glance' (1997/1903: 175) as Simmel puts it, to enable elucidation, analysis, critique? Metaphors of violent immobilization are almost invariably evoked here (as I have done) – seizure, capture, holding, grasping, arresting – as though the very attempt to bring critical attention to bear on these phenomena involves a kind of visual subjugation. Moreover, these procedures are suggested by the terms 'critique' and 'analysis' themselves; the former associated, through the meaning of its Greek root *krinein*, with separation, distinction, and judgement ('crisis' shares the same root), the latter referring to the dissection or breaking down of a phenomena into constitutive components: both terms are implicated in the distinction of a body from its surroundings. Such a distinction accompanies the primary act of attentiveness upon which the 'ground of the image' may be constituted (if we are to believe Nancy 2005), separating the object from temporal and spatial dynamics of flux and inchoateness and detaining it before the viewer. And it is this attentiveness that performs the 'fit' or affinity between the critical procedure and the images that are its object, producing what we might call the 'attentive fallacy' that informs so many theoretical approaches and methodologies that engage with images: the seemingly self-evident idea that the significance of images – and the path to understanding them – is generated through a distinct, focused encounter between a visually immobilized viewer and a discrete and equally stationary image.

The 'attentive fallacy' has not, of course, gone unnoticed or unchallenged. In *Vision and Painting* (1983) Norman Bryson charted the historical development of Western art history in terms of the opposition between an immobilizing and attentive Gaze and its suppressed other, the Glance, 'a furtive or sideways look whose attention is always elsewhere, which shifts to conceal its own existence' (1983: 94). Beyond the field of art history, others have also been concerned with issues of visual mobility and attentiveness in relation to media products. In film studies, Anne Friedberg (1993), developing Benjamin's comments on the relation between contemplation and distraction, has discussed cinema in the context of modern technologies that achieve a *perceptually* immersive ocular mobility through the *physical* immobilization of spectators' bodies and their attentive fixity within the cinematic apparatus. In media studies, John Ellis (1982) and John Caldwell (1995) have engaged in a minor disagreement over whether television is viewed in less or more attentive modes (a debate to which I shall return later on). And perhaps more than anyone else, Jonathan Crary (1992, 2002) has promoted the critical historicization of the theme of attention, and the socio-historical dialectic of visual mobility and immobilization in modern spectatorship. Central among his theses is that the subjectivization of vision in the nineteenth-century – the perception that vision had become unstable and unpredictable – also gave rise to corresponding strategies for controlling and managing the individual observer, chief among which was 'a disciplinary regime of attentiveness' (Crary 1992: 24). Last but by no means least, one could argue that the very theorization of the postmodern as an aesthetic of surface play, opposed to modernist depth and instrumentality (Jameson 1991), depends at least in part on the implied opposition between modes of distraction and attention.

Despite this growing interest in the discursive and material production of attention, the attentive fallacy continues to play an important role in implicitly structuring theory and method across disciplines that engage with visual media and images: viewers continue to be 'positioned' by singular images – which are themselves 'positioned' by the analyst's scholarly scrutiny – whose formal and semiotic minutiae are presented as crucial to their ideological, performative or moral efficacy. Working against this grain, in what follows I will take up the question of visual inattention in relation to two different media, revisiting a fictional dramatization of photographic viewing (Frosh 1995), and then developing my argument to consider pre-digital television. Through these two cases I will offer a somewhat counter-intuitive account of the morally enabling role of routinized, mediated inattention in producing what Scannell (1996) calls 'the care structure' of modern societies, especially our moral concern for the well-being and welfare of distant strangers.

The Materiality of Inattention

To make this argument I will need to consider the materiality of visual inattention. 'Inattention' functions here as a taxonomic term that covers a spectrum of visual modes, all of which have in common brief duration and low cognitive and emotional intensity, and all of which are – like attention itself – circulatory forms of connectional energy between perceiving subjects and the potential objects of their perception. It is tempting to subsume these modes under the rubric of 'affect' as it is currently being employed in the humanities: 'autonomic' bodily responses that exceed and precede conscious states of perception, cognition and emotion, pointing to a fluid, dynamic and trans-subjective 'visceral perception' that embodies potentiality and indeterminacy (Massumi 2002; Clough 2008). Yet to do so would be to lose much of the routinized, semi-conscious and determinate nature of the modes I will be exploring, as well as to embrace vitalist undercurrents that require more extensive theorization in relation to inattention than the scope of this chapter allows.

Rather, visual inattention will be explored here as a category of social-material practice. By this I mean two things, drawing on Andreas Reckwitz's (2002) productive integration of Schatzki's pratice theory and Latour's actor-network theory. First, visual inattention is clearly an *embodied* routine, organized by practical knowledge that – while rarely experienced as a cognitively conspicuous event, and only infrequently rendered into discourse – is nevertheless amenable to conscious reflection and articulation. As Reckwitz notes: 'A social practice is a regular bodily activity held together by a socially standardized way of understanding and knowing' (Reckwitz 2002: 211). We – our bodies, eyes, hands, minds – 'know' how to watch television, just as we 'know' how to flip through magazines and scan newspaper images. The first kind of materiality with which inattention is imbued is therefore that of the body's routine practical activity.

At the same time, however, inattention is a practice that is materialized in predictable and habitual situations of contact with non-human material artifacts. These artifacts are not just the 'objects' of a practice, the operative surfaces upon which it is performed or the instruments through which it is realized: they are co-agents of its social constitution and production in particular forms under 'known' conditions: 'The things handled in a social practice must be treated as necessary com- ponents for a practice to be "practiced" … When human agents have developed certain forms of know-how concerning certain things, these things "materalize" or "incorporate" this knowledge within the practice … Things are "materialized understanding"' (Reckwitz 2002: 212). In the discussion below the 'things' that act as 'materialized understandings' of a social practice (in interaction with our bodies) are representational media technologies (photographs and photographic albums; television images and television screens). And the practices they materialize and routinely co-constitute – among others – are modes of visual inattention.

This double significance of 'materiality' – the practical materiality of the body in routine intersection with non-human artifacts – also suggests a double negation. The first negation is that materiality itself is often defined in the humanistic disciplines most concerned with representation as the exterior of meaning: the non-representational substrate, the physical or functional residue that remains once a sign has been 'de-semanticized' and stripped of its semiotic kernel. The material, in this hermeneutic schema, is the body-vehicle that carries meaning but is external and supplementary to it: it is the result of the evacuation of representation.[1]

Hans Ulrich Gumbrecht attempts to challenge this negation – and the matter-spirit dualism on which it is ultimately based – by defining 'materialities of communication' as 'the totality of phenomenon contributing to the constitution of meaning without being meaning themselves' (1994: 398). Yet this attempt, laudable as it is within the hermeneutic parameters of the humanities, risks simply inverting the matter-spirit dualism in a manner all too familiar to social scientists, privileging the extra-hermeneutic category of 'material conditions' (including variables associated not just with the economic 'base' but with 'social structure' more broadly) as 'underlying' *determinants* of meaning (Brown 2010). So the humanistic negation of materiality needs to be met not with its own inversion (a negation of the negation) but with a continual questioning and crossing of the matter-meaning distinction itself: no matter without meaning, and no meaning without matter.

Visual inattention offers a particularly rich field for such questioning and crossing. On the one hand it is a socio-material practice that integrates and emerges from a network of interacting forces: bodies in action, spatial and temporal frameworks, technological artifacts and forms of representation. On the other hand, since these forces repeatedly construct it, then 'inattention' needs to be conceived of in terms that resist its own designation as *in*-attention: *as the negation of the very connective conditions* (focus, concentration) *thought to enable the meaning-carrying unit – the image-sign – to signify in the first place.* Visual inattention, I will hope to show, is a significant material practice not because it precedes, negates or is external to representation, but because it participates in the routine modulation of alternate connectional energies, which in turn are productive of alternate representational forms that are not reducible to the discrete, attentively viewed image. It is these alternate energies that, I argue, underpin our everyday moral concerns for distant others.

---

1   This is of course a key theme in Derrida's early critique of 'logocentric' accounts of meaning, language and writing (1976).

Photography: 'They're all the Same!'

Wayne Wang's 1995 film *Smoke* contains a beautiful and extremely illuminating cinematic construction of photographic viewing. In a scene near the beginning of the film Auggie shows Paul his photograph albums. Each page of these indexed and dated volumes contains six white-framed black-and-white images; each image was taken on a different day, at precisely 8 o'clock in the morning, from exactly the same spot (the corner of 3rd Street and Seventh Avenue in New York), at the same angle and of the same field of view (the street corner in front of Auggie's tobacco shop), every day of the year. Auggie explains that he can never take a holiday because he has to be on 'his spot' every morning to take a picture, in all weathers: 'Sort of like the postman'.

As he flicks through these albums, Paul is amazed by the sheer excess of identical photographs (more than 4,000 of them), and the self-defeating futility of Auggie's enterprise.

> 'They're all the same!' Paul observes.

> Auggie responds: 'They're all the same, but each one is different from every other one. You've got your bright mornings and your dark mornings. You've got your summer light and your autumn light. You've got your weekdays and your weekends … The earth revolves around the sun, and every day the light from the sun hits the earth at a different angle.'

> Paul needs to slow down, Auggie suggests: 'You're going too fast. You're hardly even looking at the pictures.'

He needs to contemplate every image separately, if he is to see that in fact each image – and the momentary reality it depicts – is unique. Though initially sceptical, Paul begins to follow this advice, focusing his attention on individual photographs, which are made to fill the screen in a cinematic re-presentation of his activity. Finally, he comes across an image of his dead wife Ellen, and he cries. The scene then suddenly shifts to a shot of Auggie standing behind his camera in his 'spot' opposite his store in the morning: he looks at his watch, presses the button on his camera to take a picture, then looks through the camera's viewfinder, and makes a note in his notebook.

There are many things one could say about this sequence and in particular about the habits of looking it exhibits and performs. Most obviously, perhaps, it suggests that there is a connection between photographic resemblance and the material (bodily) mode and context of viewing, specifically the duration, mobility and concentration of the look, as well as the material (artifactual) method of displaying the photographs. I use the term 'photographic resemblance' in a double sense here. On the one hand it designates the production of iconic similarity between photographic images, the creation of a typological identity that allows one to say of a group of photographs 'They're all the same'. On the other hand, it refers to the indexical singularity of each and every photograph: the photograph's semiotic status as 'an imprint or transfer off the real', as what Barthes calls the 'absolute Particular, the sovereign Contingency' (1984: 4), a unique and unclassifiable material trace of its referent at an unrepeatable moment in time.

Mobility, duration, intensity, mode of presentation; this scene demonstrates that these material factors can radically alter our sense of the sameness and dissimilarity of images. As the iconic similarity between photographs is made more conspicuous, so their fidelity to particular referents is diminished; as the indexical singularity of each image is emphasized, so their correspondence to one another decreases and it becomes erroneous to say that they are 'all the same'. Paul's original inattentiveness takes its cue in part from Auggie's decision to display the photographs in albums: he is merely performing the everyday visual-tactile practice of image-scanning and page-flicking by which we work our way through newspapers, magazines and, indeed, family photo-albums (especially other people's). Auggie actively has to intervene in order to overcome this embodied-technical routine of glancing, urging Paul to slow down and pay attention. In the process he not only raises the act of looking to the level of consciousness, de-routinizing vision; he also reconstitutes it around an altogether different set of bodily practices and technical instruments, the exhibition spaces (museum, gallery) and prolonged gaze appropriate to the proper reception of art.

By associating immobilization with the 'truth' of photography as against the viewer's roving, scanning, distracted glance, this scene shows how the problem of visual movement becomes particularly aggravated in the case of photography.[2] For thanks to the almost infinite reproducibility of photographic images, the multitude of contexts in which they can be seen and the variety of material vehicles on which they can be displayed, as well as their consequent proliferation across social and cultural realms, photographs seem to be integrated – more seamlessly perhaps than other representations – into a total, fluctuating environment in which the individual image loses its singular claims on the viewer's attentive gaze. As Victor Burgin argues: 'It is therefore not an arbitrary fact that photographs are deployed so that we need not look at them for long, and so that, almost invariably, another photograph is always already in position to receive the displaced look' (1982: 143).

Ultimately, Auggie's need to intervene in order to alter the way Paul looks at his photographs raises the suspicion that viewers are indifferent to the photographs that they see. This indifference is not one of boredom, ennui or alienation, but is a material practice: a routinization of corporeal and perceptual connective energies, both tactile and visual, that produces sameness from movement in a context

---

2   The vision of the cinema spectator watching this scene is in fact heavily controlled by filming techniques (close-ups of particular photographs, fade-in and fade-out, musical enhancement) and the cinematic apparatus (large screen size, darkened auditorium, relatively immobilized spectators) to reproduce the mode of attentive looking favoured by Auggie.

characterized by the superfluity of representations and perceptual stimuli. In other words, against the hierarchal privileging of singularity and visual attentiveness as key characteristics of photographic significance, it is actually the qualities of indifference, sameness and visual displacement that routinely serve as the ground for experiencing photography's way of showing the world.

The indifference of the viewer is strikingly parallelled in this scene by the indifference of the image-producer to the detailed representational construction of his own images. One of the most perplexing things about Auggie's albums in *Smoke* is that they represent an oeuvre based entirely on a mechanized and automated indexical process. Once Auggie has set the parameters for all the photographs (time, place, field of view), his presence at the scene becomes irrelevant to the actual production of the pictures; the same results could be obtained from a computer-operated camera. Each photograph in Auggie's albums is the result of an arbitrary, uniform and inexorable predetermination (to take a picture at exactly the same time of exactly the same scene, day after day) that, once set in motion, prevents meaningful, context-sensitive human intervention. He is so removed from influencing the content of his images that we see him check the camera's viewfinder only *after* he has actually taken the picture. In a manner that fully echoes Fox Talbot's (1844–1846) famous description of photography as the 'pencil of nature', reality seems to reproduce its own image through the agency of light and a mechanical device. As Baker notes, 'in any photograph, the object depicted has impressed itself through the agency of light and chemicals alone, inscribing a referential excess beyond the control of the creator of any given image' (1996: 75). Auggie simply takes this breach between image and author to its extreme, voluntarily maintaining his indifference to the particular content of any particular image.

This referential excess of the photograph – what Berger calls its 'weak intentionality' (1982: 90) – is closely related to the indifference of viewers. No longer asked to reproduce the primary perception of an artist, the viewer's eyes are confronted by a reality which appears routinely to recreate itself as a matter of course.[3] Rather than a series of singular images attracting and returning the gaze of the viewer, photographs aggregate into a kind of fluctuating backdrop which is 'taken in' rather than intensively viewed, producing the seemingly unmotivated autopoetic unfolding of a depicted world.

So if photography enjoys indifferent relations both with the world it depicts and the viewers who behold it, how might these possess anything resembling moral significance? Much of the answer to this will be developed in the discussion of television and in my concluding remarks, but for the moment a few intermediate points are required.

---

3   It is this 'indifference' of the photograph to the organizing perception of either artist of viewer that leads Martin Jay (1988) to classify photography as chiefly corresponding to the 'scopic regime' of 'The Art of Describing' rather than to 'Cartesian Perspectivalism'.

First, photographic indexicality – by virtue of its weak intentionality – is radically inclusive (this is, of course, another central meaning of the word 'indifferent'): it registers both the principal subject and the extraneous detail, this being of course a key condition of Barthes' *punctum*. John Ellis makes this one of the main themes of his claim that mass media have created a new generalized mode of relating to the world, a new form of witnessing, and what is true for photography is even more apposite for cinema and television:

> The most astonishing thing [about the experience of early cinema] was that everything in the picture moved, "even the leaves on the trees" as one observer put it … It was the sudden ability to witness the incidentals of life just as they were that produced the effect of witness … Photography had the ability to capture everything that lay in front of the lens. The film camera was able to give it all motion. Together they introduced the audiences of a century ago to a new potential to witness events and phenomena in the world around them (2000: 19–20).

Auggie of course takes this to its extreme; by surrendering virtually all control of detail he makes the accidental and the incidental decisive elements in the production of each image. In this he is representative of the witnessing effect of audiovisual mass media more broadly, generalizing witnessing from this specified event and that particular experience to the perpetual presentation of worlds.

Second, while the photograph is the result of a mechanical process, it is nevertheless 'programmed'. There is always a *generic* intentionality in operation behind it – cameras, even automated ones, do not appear of their own accord – and this intentionality can be interpreted by viewers. Auggie's generic intention is documentary as well as proprietary: to report on his 'spot', his little corner of the world. He claims that in order to 'get' his project one needs to slow down and appreciate the particularity and singularity of each image, and the uniqueness of each individual represented. But this only makes sense within the framework of an overall project of indifferent world-presentation, whereby one 'inattentively' judges these photographs to be 'all the same' in the sense that they indifferently record *the same world* – an actual world of contingency shared by the viewer – and the individuals appear simply by virtue of the fact that they have wandered into it at that moment. It is the intentional ground of homogeneity and identity ('this spot') that acts as the fundamental framework within which specific encounters with particular images and particular individuals then become possible (Ellen, says Auggie, just happened to be passing on her way to work).

This allows photographic indifference to be revolutionary in another sense, in that it gives a central role to *indexical anonymity*: you can be reproduced through a causal and physical process in an image in exquisite detail, in an existentially unrepeatable field of space and time – it is definitively you, there, then – and yet remain unknown to your photographer and to your viewers: this occurs most conspicuously in particular 'iconic' images (Hariman and Lucaites 2007) where anonymous individuals – such as Dorothea Lange's 'Migrant Mother' – can come to stand in for larger collectivities and events which are not depicted. Moreover, photography extends this gift to anyone, however irrelevant to the putative central topic of the image, simply by virtue of their entering the camera's field of view at the moment the photograph is taken. It therefore speaks *simultaneously* to social generality – you appear as an unidentified anyone – and to unique individuality – it is you, a particular someone, and no other, in that place and at that time. Ellen is singled out by Paul and by Auggie, but she is in fact no different to all those other anyones who populate Auggie's images and who are also always potentially someone in particular to a viewer somewhere.

Finally, what Auggie's project demonstrates in microcosm is photographic indifference as a relentless, routinized, bureaucratic system of same-world disclosure. It is no accident that Auggie cannot take a holiday, that he has to be on his spot at 8am every morning, rain or shine, or that he compares himself to the post man, thereby invoking the first modern mass communication bureacracy. And, like mass media, Auggie's project is also relentlessly bureaucratic and archival: at one point in the scene described earlier Paul chuckles sardonically on being presented with the next in a long line of carefully labelled and dated photograph albums. Auggie's photographic indifference is perpetual and regulated, mundane and everyday. It happens irrespective of whether or not there is a particular individual or event judged a prioi as worth attending to, because it does not convey specific information or experience but presents us with an inclusive chronotope, an unfolding world.

Television: Are you Looking at Me?

So far I have linked the inattentive character of much habitual viewing of photographs – dramatized in the scene from *Smoke* – with the inclusiveness, contingency and indifference of photographic indexicality: this is an example of the way material practices are bound up with, rather than exterior to, representational forms. The iconic similarity attributed inattentively to photographs as a routinely encountered visual environment complements the same-world disclosure that emerges, paradoxically, from their indexical nature. Television is, of course, a different medium, materially and socially constructed with different affordances, expressive constraints and possibilities, and contexts of engagement with viewers. Inattention, in the case of television, is chiefly associated with what later came to be called 'glance theory': the thesis that viewers' engagement with the medium is largely characterized by a lack of visual concentration. John Ellis made probably the best known and most persuasive case for this idea in *Visible Fictions* (1982): the small size of the television screen, coupled with the distracting domestic circumstances of viewing, mean that 'TV is not usually the only thing going on, sometimes it is not even the principal thing. TV is treated casually rather than concentratedly' (128). 'Glance theory' has been challenged over the years, most comprehensively and engagingly by John Caldwell, who criticizes the generality of the claims made for casual viewing and Ellis' neglect of the hard material and stylistic labour invested by programmers in attracting viewer attention as a matter of narrative and commercial necessity. 'The viewer is not always, nor inherently, distracted.' Caldwell concludes: 'Theorists should not jump to theoretical conclusions just because there is an ironing board in the room' (1995: 27).

In Ellis' defence his argument was published some thirteen years prior to Caldwell's critique and with reference to British rather than US television (Caldwell acknowledges as much: 365, note 57) – though it was also controversial when first published (Winston 1984). Ellis also works from a naturalistic phenomenology of the viewing experience, whereas Caldwell focuses on the logics, technologies and practices of television production, which are concerned with attracting and maintaining viewer attention in the context of multi-channel competition and a relatively weak apparatus of attentiveness (compared to cinema). Both, of course, are partly justified in their claims: Ellis in arguing that being distracted is more likely and acceptable, is even a horizon of expectation, when watching television at home than when going to the cinema; Caldwell in insisting that television systematically employs numerous visual techniques – many of them decidedly cinematic – for attracting and holding attention.

However, the terms of this disagreement are largely conditioned by an emphasis on television as a representational device: a machine for making images. Since, however, visual inattention primarily describes forms of connectional energy, it makes sense to explore it as a socio-material practice – along with its moral potentialities – in relation to another well-documented feature of television: its nature as a medium of connectivity between viewers and the social whole, in particular through conventions of temporal simultaneity and 'para-social' interaction (Horton and Wohl 1956).

Television introduces a socially novel form of visual connectivity: non-reciprocal face-to-face communication. Television allows one to be 'face-to-face' with another person and *not* pay them any attention. The appearance of another's face on the screen, even when accompanied by direct verbal address, the illusion of physical proximity and temporal simultaneity (live transmission), is an indication of their non-presence at the location of viewing. One can maintain an attitude of utter indifference, even when apparently being directly spoken to, ignoring both their face and their words. Crucially, this non-reciprocity is itself mutual, a result of the systemic organization of technologies of 'mediated quasi-interaction' (Thompson 1995). The actor, the anchor in the studio, the politician speaking to camera: all are impervious to our efforts at interruption, and are entirely ignorant of any lack of attention on our part.

Though the rhetorics of direct address combined with the logics of mediated non-presence have long been analyzed, in a variety of ways (see Horton and Wohl 1956; Tolson 1996; Scannell 1996), the fundamental moral significance of this mutual indifference should not be under-estimated. What positive moral relations could possibly be produced by the ability to ignore others? Non-reciprocal face-to-face interaction seems to resemble the kinds of indifferent visual encounters that characterize actual physical proximity and co-presence in modern public spaces – such as Simmel's analysis of mental life in the metropolis and Goffman's 'civil inattention' (1963; see also Koch 1995). To its potential discredit, perhaps, television appears to introduce such non-attentive mechanisms for apprehending unknown others into the home.[4] This domestication of inattentive relations with strangers invites a severe critique: surely it is an extension of alienated public interaction norms into the private sphere, de-ethicalizing others by turning them into mere background figures (Bauman 1990)? Critiques similar to this are common in media studies. The non-reciprocity of media is transformed into a figure of moral distance between viewer and viewed: the lack of congruence between those depicted and the local 'relevance structures' of 'direct experience' in the lives of viewers hinders the nurturing of care and responsive action (Tomlinson 1999). Audiovisual media technologies create non-reciprocal non-encounters between viewers and viewed and are perceived to insulate the viewer from ethical responsibility to those represented on the screen: the screen functions as a window but also as a barrier, allowing us 'to maintain a considerable distance from what we see and thus to acquire an anaesthetised form of knowledge' (Morley 2000: 184).

There is, however, an alterative assessment: that non-reciprocal face-to-face communication allows the faces of strangers to appear in the sphere of intimacy without creating alarm or triggering a defensively hostile response (because these faces can be safely ignored). This more optimistic view sees televisual non-reciprocity as an extension, domestication and enhancement of 'civil inattention' in public contexts, but reaches an entirely different conclusion concerning its moral potentialities. It assumes that encounters with multiple strangers in public are potentially a recipe not for dialogue but for the defensive expression of hostility and aggression, and that encounters with the uninvited faces of strangers in the private sphere – on one's own personal territory – are potentially even more explosive. In this context, civil inattention, epitomized by the way individuals briefly glance across the eyes and faces of physically proximate others in a public space, is 'perhaps the slightest of interpersonal rituals, yet one that constantly regulates the social intercourse of persons in our society' (Goffman 1963: 84). Its great moral importance is invested in the equal connection that it accords to anonymous others, recognizing their co-presence as non-threatening, and therefore as unworthy of particular attention or interest: 'By according civil inattention, the individual implies that he has no reason to suspect the intentions of the others present and no reason to fear the others, be hostile to them, or wish to avoid them' (1963: 84). What one would wish to avoid, in other words, is uncivil attention: the

---

4   Of course, there are important pragmatic differences between face-to-face civil inattention and non-reciprocal televisual face-to-face communication, not least the fact that being looked at by a co-present stranger in public is conventionally interpretable as a hostile act. This potential interpretation of personally directed hostility is diffused in television viewing.

kind of attention lavished upon strangers in public places by children who have yet to learn that it is rude to stare or point at those who look different, or by security forces in the days following terrorist attacks, when the non-hostility of strangers is a matter of deadly uncertainty (Frosh 2007).

Television's performance of non-reciprocal face-to-face communication is therefore morally significant as a domesticator of civil inattention. It allows the faces and voices of strangers to become visible in the sphere of intimacy without their very appearance in the home creating alarm or triggering a defensively hostile response. The multiplicity of ignorable faces on the television screen makes a diverse range of strangers constantly quasi-available, often unattended to but always in attendance like the medium itself, potentially connected to viewers who can chose to engage with them or not, and who need not feel threatened by them. By neutralizing the potential threat of the stranger, televisual indifference ensures that vast numbers of unknown others become liveable-with in our homes and personal lives, co-habitable over the long term.

Being with the Composite Image

The unthreatening co-habitability of multiple strangers made possible by televisual non-reciprocity nevertheless seems to command a high price: responsiveness to the particularity of an individual other. The lack of attention paid to the faces of others, apparently looking at and speaking directly to us, reduces the intensity of our encounter, but it also undermines the perception of their singularity. More to the point, the lack of attention built into the very structure of televisual non-reciprocity is reinforced by another feature at odds with the immobilizing assumptions of the attentive fallacy: the ephemerality of most strangers on television.

Ordinary individuals on television are transient. They appear for a few seconds before being replaced by other images, other faces and bodies. Indeed, it is their very transience on screen that helps to mark them as 'ordinary people' who hail from the shared 'real life' reference world beyond television, as opposed to the corporeal familiarity of celebrities, journalists, actors and presenters whose repeated appearance on television situates them within 'the media' as a seemingly autonomous realm (Couldry 2000). Unlike characters in novels, or fictional characters in television serials and soap operas, we do not have time to 'get to know them' and to develop feelings of intimacy with their personalities or imaginary emotional attachments to them. They are somewhat similar to strangers encountered in public places: we tend to overlook them rather than look at them, and can rarely recall them in any detail as unique individuals.

Forgettable and largely unknowable as specific individuals, these transients usually undergo a process of generalization. There are two intertwined mechanisms shaping this process. The first is the particular semiotic combination of indexical, iconic and symbolic dimensions of modern visual and audio-visual media already discussed in the case of photography and the scene from *Smoke*, which makes possible the indexical anonymity of individuals. Yet, as with photography, so with television there seems to be a price to pay for turning the individual into a metonymic emblem of broader, undepicted populations and events: the erasure of the individual as a singular being, a memorable, unique person who can form the basis for a concrete appeal to moral action among audiences. The move from index to emblem appears to obliterate the former.

Unlike photographs, however, television images do not stand still: or rather, their circulation and mobility is technically automated rather than a product of visual, corporeal and material practices performed upon still pictures. Hence the second mechanism shaping the generalization of depicted strangers on television has to do with the incessant temporal flow of images, sounds and texts that is a default technical characteristic of the medium's mode of display. It is not only the viewer's gaze or attention that wanders: the medium is based upon the succession of images. Though there is certainly an economy of attentiveness and familiarity created by the repetitive cycles of television programming (daily, weekly and seasonal schedules; reruns: Ellis 2000, Kompare 2005), as well as a less predictable recycling and circulation of particular images, the individuality of strange faces and bodies is affected by both the viewer's movement across stimuli and by a key temporal dimension of the medium itself. In this context images of individuals can be generalized as *units of aggregation* across texts and genres, overlaid with multiple others judged to be similar in an incessantly accumulating image-stream.

The relationship between the generalized image and the singular individual, then, is as much a matter of expansion and accumulation as it is of erasure. It is the erasure of the *boundaries* of the particular individual, not the annihilation of the individual per se, that is accomplished. If most ordinary individual strangers on television are viewed indifferently and transiently, in routine, unremarkable, non-hostile encounters, then their constant and cumulative presence within the home is a significant historical accomplishment. It produces for individuals in their intimate spaces a serial aggregate of the human figure as a shared 'condition', an instrument of similarity and interconnection, an interminably fluctuating and ever-present composite image: perhaps, even, the 'face' of humanity itself. It is only in so far as they are recognizable constituents of this serial aggregate (to which they are in turn added) that the ethical appeal of particular images and stories of specified individuals can be produced.

The term 'composite image' has a troubling history: its association with social Darwinism and the racist and eugenicist missions of nineteenth-century practitioners such as Francis Galton behooves us to treat it with a degree of caution (Sekula 1989). But it is also important to recall Carlo Ginzburg's (2004) description of the composite image as a kind of 'family resemblance', which in aggregating a multitude of individual faces produces not uniformity and homogeneity but a porous, overlapping and fluid identity that is in excess of the disciplinary frameworks in which it was developed (indeed, in Sekula's well-known account this – largely indexical – excess underpinned the failure of the composite image as a useful tool of social engineering). To be com/posite is to be situated with, alongside, in company with. The non-threatening serial aggregate of televised strangers produces a visual figure for the composition of multiple selves and others, for our constitutive superimposition and intermingling as *both* singular *and* similar beings, for our ever-emerging and changeable 'human' commonality. To go out on a limb here, it is a figure of what Nancy (2000) would call our singular-plural 'compearance', our foundational situation of 'being-with'.

Of course, the status of 'the human' as such, and as the product of a putatively inclusive visual 'language' has been the subject of much criticism. Especially well-known is Barthes' essay 'The Great Family of Man' in *Mythologies* (1993/1972) in which Barthes attacks both the universality of the photographic medium and the universality of 'man' as myths whose effects are to dehistoricize and naturalize the current social order. Against this critique I would argue that production of an unthreatening, perpetually aggregating composite image of strangers nevertheless socially institutionalizes a space in which each individual's extended and abstract relations of similarity with distant, unfamiliar others become definable under the category of 'the human'. This space is moral rather than ethical, in the sense that it guides our 'thin relations', our behaviour 'toward those to whom we are related just by virtue of their being fellow human beings' (Margalit 2002: 37), rather than our 'thick relations', our behavior toward those with whom we enjoy strong social bonds (family, friends etc.). Inattentive and indifferent relations to a fluctuating serial figure of 'the human' act as an unexciting but central routinizing procedure for the moralizing of strangers as human: based on a logic of superficial iconic similarity – of sufficient substitutability between individuals (they're all the same!) that nevertheless preserves their singularity – televisual indifference, along with the indifferent same-world unfolding of photography, means that 'others' are always already other *people*. As Nancy observes: '"People" clearly states that we are all precisely *people*, that is, indistinctly persons, humans, all of a common "kind" … This existence can only be grasped in the paradoxical simultaneity of togetherness (anonymous, confused and indeed massive) and disseminated singularity (these or those "people(s)", or "a guy", "a girl", "a kid"' (2000: 7).

Conclusion: Totemic Inattention

Photography and television are two distinct media, and although my account of their inattentive and indifferent relations to their viewers highlights commonalities between them, it is to a degree an abstraction concerned with typifications of viewer behaviour (glancing, non-reciprocal connectivity) and technical affordances (modes of display, image-transience), rather than an analysis of how inattention is performed and embedded in the lives of empirical viewers. While I have described visual inattention as an embodied and technically realized socio-material practice, it is nevertheless a 'genus', a taxonomic category distinct from the specificities of viewers' life-worlds. In its generality and distance from 'actual' data my discussion has constituted a second-order materialism, one level of the multiple 'orders of materiality' that Brown describes as constituting an (unrealizable) 'ideal materialism' (2010: 59).

There is, then, a formal affinity between the aggregative consequences of inattention that I have teased out and the aggregative method of analysis that I have employed. At worst, my analysis is the product of an insufficiently attentive approach, one insensitive to the myriad practices, affects, structures and relations that terms like 'indifference' both designate and conflate. And perhaps it also echoes Holly's claim – cited at the beginning of this article – that there is 'a productive correspondence of rhetorical ideologies between image and text. Representational practices encoded in works of art continue to be encoded within their commentaries' (1990: 385), provided that we gloss 'works of art' inclusively to refer to media and their products in general.

I would venture, however, to argue that such a productive correspondence in this case proceeds 'totemically', basing my claim on the revitalization and refiguration of the concept of the totem in recent work by W.J.T Mitchell (2005). It is totemic in at least one obvious sense: it has detailed the routine production and performance of a totemic being – the composite image of 'the human' – 'where the species-being of the individual is "crystallized", as it were, and rendered as a kind of concrete universal' (Mitchell 2005: 178; note 22). This totemic being is unusual in terms of traditional anthropological understandings of the concept since its form is not that of an animal or natural object, but rather that of the human face and figure. Or rather, perhaps, this difference bespeaks the givenness of an immediate perceptible universe populated more by images of distant human individuals and multitudes from whom we take our measure than by facets of the non-human 'natural' world. The culture of media and image-saturation has become, ultimately, 'second nature' to the extent that 'humanity' has become auto-totemic: it is its own lived clan-symbol.[5]

The composite-image produced by inattentive viewing, image-mobility and media-ubiquity is, like the totem, *animated*. It is fluid, unfixed, constantly modulated not only by the individual and social materialities of viewing, but by the additive and cumulative dynamics of contemporary media themselves. This means that the composite image of the human is constituted through incessant transformations. The sameness it produces by accumulation and aggregation are 'repetitions' that enable recognition of multiple others *as such*, as definable figures against the flux, and as always prior to and beyond singularity: 'See statis, see station, as a special case of movement (a special case of reiterative movement: that allowing recognition)' (Massumi 2002: 66). Having said this, there are undoubtedly political and economic structures to this accumulation: certain populations enjoy considerable representation (in both political and semiotic senses of the word) and recognition while others remain undepicted and abject –

---

5   The totemic power of the human face – and its sickening violation – can be glimpsed in the image of the totalitarian future described in Orwell's 1984: a boot stamping on a human face, forever.

unrecognized as subjects of discourse (Butler 1993). Hence the moral productivity of inattentive viewing also needs to be considered relationally, in connection to forms of institutionalized media attention that are usually highly centralized (notwithstanding the development of digital technologies and networked media).

This sense of composite image as a human totem remains entirely susceptible to Barthes' ideological critique of the myth of the 'Family of Man' mentioned earlier, as a naturalization of the particular social and historical conditions that produce it. But to make this critique is – again in Mitchell's terms – to transvalue the totem into an idol, an image of (false) veneration and power that invites iconoclastic deconstruction, and which usually involves its replacement by another idol, often an 'idol of the mind' (Mitchell 2005: 189, borrowing from Bacon). In Barthes' case the new, substitute idol can be found in the explicit scientism of the semiological method, one in which Barthes – as 'mythologist' – needs of necessity to distinguish himself from a mere everyday 'myth-consumer': 'When a myth reaches the entire community, it is from the latter that the mythologist must become estranged if he wants to liberate the myth ... The mythologist is condemned to live in a theoretical sociality' (1993: 156–7). However, in tracing the correspondence between aggregations and frameworks of viewer inattention, as well as the ubiquity, mobility and indifference of images in modern media, another feature of the totem has been engaged, one that partakes of a sociality both theoretical *and* material: that the totem embodies, symbolizes and performs relationality.

By this I mean not only that the totem explicitly expresses and performs social bonds (whereas idols and fetishes tend to disguise or replace them), though it is of more than mere passing interest that 'totem' literally means 'relative of mine' in the Ojibway language from it was originally taken (Mitchell 2005: 98). Rather it is that it operates in the cumulative and connective tenor of 'both and' rather than the analytical binary of 'either/or'. The inattentive mode of viewing photographs is connective rather than discriminating; non-reciprocal face-to-face communication enables a plurality of weak ties with ever increasing numbers of unfamiliar strangers rather than exclusive commitment to particular, identifiable individuals. The composite image itself is *both* indexical *and* emblematic, singular and general, someone and anyone, change and repetition, concrete particular and abstract universal. It links the habitual operations of inattentive and indifferent modes of viewing, performed by individuals in their everyday spaces and lives, to the cohabitation of extended worlds with many anonymous others. In the process it produces 'humanity' as both a background figure in perpetual attendance – a stranger-companion – and as the ultimate social aggregation.

References

Abercrombie, N. and Longhurst, B. (1998), *Audiences: A Sociological Theory of Performance and Imagination* (London: Sage).

Baker, G. (1996), 'Photography Between Narrativity and Stasis: August Sander, Degeneration and the Decay of the Portrait', October, 76: 72–113.

Barthes, R. (1993/1972), *Mythologies* (London: Vintage).

Barthes, R. (1984), *Camera Lucida: Reflections on Photography* (London: Fontana).

Bauman, Z. (1990), 'Effacing the Face: On the Social Management of Moral Proximity', *Theory, Culture and Society*, 7: 5–38.

Berger, J. (1982), 'Appearances' in J. Berger and J. Mohr (eds) *Another Way of Telling* (New York: Pantheon Books), 85–100.

Boltanski, L. (1999), *Distant Suffering: Morality, Media and Politics* (Cambridge: Cambridge University Press).

Brown, B. (2010), 'Materiality', in W.J.T. Mitchell and M.B.N. Hansen (eds) *Critical Terms for Media Studies* (Chicago: Chicago University Press), 49–63.

Burgin, V. (1982), 'Photography, Phantasy, Function' in V. Burgin (ed.) *Thinking Photography* (London: Macmillan Education).

Butler, J. (1993), *Bodies that Matter: On the Discursive Limits of 'Sex'* (London: Routledge).

Bryson, N. (1983), *Vision and Painting: The Logic of the Gaze* (New Haven: Yale University Press).

Caldwell, J. (1995), *Televisuality: Style, Crisis and Authority in American Television* (New Brunswick: Rutgers University Press).

Chouliaraki, L. (2006), *The Spectatorship of Suffering* (London: Sage).

Clough, P. (2008), 'The Affective Turn: Political Economy, Biomedia and Bodies', *Theory, Culture and Society*, 25: 1, 1–22.

Couldry, N. (2000), *The Place of Media Power: Pilgrims and Witnesses of the Media Age* (London: Routledge).

Crary, J. (1992), *Techniques of the Observer: On Vision and Modernity in the Nineteenth Century* (Cambridge: MIT Press).

Crary, J. (2002), *Suspensions of Perception: Attention, Spectacle and Modern Culture* (Cambridge: MIT Press).

Dayan, D. (2009), 'Sharing and Showing: Television as Monstration', *The Annals of the American Academy of Political and Social Science*, 625: 19–31.

Derrida, J. (1976), *Of Grammatology* (Baltimore: The Johns Hopkins University Press).

Ellis, J. (1982), *Visible Fictions* (London: Routledge).

Ellis, J. (2000), *Seeing Things: Television in the Age of Uncertainty* (London: I. B. Tauris).

Friedberg, A. (1993), *Window Shopping: Cinema and the Postmodern* (Berkeley: University of California Press).

Frosh, P. (2003), *The Image Factory: Consumer Culture, Photography and the Visual Content Industry* (Oxford: Berg).

Frosh, P. (2007), 'Penetrating Markets, Fortifying Fences: Advertising, Consumption and Violent National Conflict', *Public Culture*, 19: 3, 461–482.

Ginzburg, C. (2004), 'Family Resemblances and Family Trees: Two Cognitive Metaphors', *Critical Inquiry*, 30: 537–556.

Goffman, E. (1963), *Behavior in Public Places: Notes on the Social Organization of Gatherings* (New York: The Free Press).

Gumbrecth, H.U. (1994), 'A Farewell to Interpretation' in H.U. Gumbrecht and K. Ludwig Pfeiffer (eds) *Materialities of Communication* (Stanford: Stanford University Press), 389–402.

Hariman, R. and Lucaites, J. (2007), *No Caption Needed: Iconic Photographs, Public Culture and Liberal Democracy* (Chicago: University of Chicago Press).

Holly, M.A. (1995), 'Past Looking' in S. Melville and B. Readings (eds) *Vision and Textuality* (London and New York: Macmillan).

Horton, D. and Wohl, R. (1956), 'Mass Communication and Para-social Interaction: Observations on Intimacy at a Distance', *Psychiatry*, 19: 215–29.

Jameson, F. (1991), *Postmodernism, or the Cultural Logic of Late Capitalism* (Durham: Duke University Press).

Jay, M. (1988), 'Scopic Regimes of Modernity' in H. Foster (ed.) *Vision and Visuality, Dia Art Foundation Discussions in Contemporary Culture* (Seattle: Bay Press), 2–23.

Koch, G. (1995), 'Nähe und Distanz: Face-to-face Kommunikation in der Moderne' in G. Koch (Hg.) *Auge und Affekt* (Frankfurt: S.Fischer Verlag), 272–291.

Kompare, D. (2005), *Rerun Nation: How Repeats Invented American Television* (London: Routledge).

Margalit, A. (2002), *The Ethics of Memory* (Cambridge: Harvard University Press).

Massumi, B. (2002), *Parables for the Virtual: Movement, Affect, Sensation* (Durham: Duke University Press).

Mitchell, W.J.T. (2005), *What Do Pictures Want? The Lives and Loves of Images* (Chicago: University of Chicago Press).

Moeller, S. (1999), *Compassion Fatigue: How the Media Sell Disease, Famine, War and Death* (London: Routledge).

Morley, D. (2000), *Home Territories: Media, Mobility and Identity* (London: Routledge).

Nancy, J-L. (2000), *Being Singular Plural* (Stanford: Stanford University Press).

Nancy, J-L. (2005), *The Ground of the Image* (New York: Fordham University Press).

Reckwitz, A. (2002), 'The Status of the "Material" in Theories of Culture: From "Social Structure" to "Artefacts"', *Journal for the Theory of Social Behaviour*, 32: 2, 195–217.

Scannell, P. (1996), *Radio, Television and Modern Life* (Oxford: Blackwell).

Sekula, A. (1989), 'The Body and the Archive' in R. Bolton (ed.) *The Contest of Meaning: Critical Histories of Photography* (Cambridge: MIT Press), 342–388.

Simmel, G. (1997/1903), 'The Metropolis and Mental Life', in D. Frisby and M. Featherstone (eds) *Simmel on Culture* (London: Sage), 174–185.

Slovic, P. (2007), '" If I look at the Mass I will Never Act": Psychic Numbing and Genocide', *Judgment and Decision Making*, 2: 2, 79–95.

Thompson, J.B. (1995), *The Media and Modernity: A Social Theory of the Media* (Cambridge: Polity Press).

Tolson, A. (1996), *Mediations: Text and Discourse in Media Studies* (London: Arnold).

Tomlinson, J. (1999), *Globalization and Culture* (Chicago: The University of Chicago Press).

Winston, B. (1984), 'Television at a Glance', *Quarterly Review of Film Studies*, 9: 3, 256–61.

Index

Page numbers in *italics* refer to figures.

natural environment
  botanical images, Caribbean 26–7
  theoretical perspectives 109–11
  time and decay 61–2
  see also Loweswater, Cumbria
Newman, O. 145
Nye, D. 21

Orlikowski, W. 135
"Orthostatic Tolerance" (Strachan) 32–3

paintings 87–8, 173
  Cézanne 98–100, 103
parks 97, 101, *102*
Parliament buildings
  London 78–83
  New Dehli 76–8
Pauli, L. 69
Pertchik, H. and Pertchik, B. 26–7
Pfahl, J. 61–2
phantom midge larva (*Chaoborus*) 119,
  126–7, 127–8, 129
photography 20–21, 61–2, 94, 149–50
  industrial landscapes 67–71
  shipbreaking 62–7, 69–71, *72*
  viewing 176–80
  vs paintings/drawings 99, 100
politics
  of justice 164–6
  of melancholia 162–4
"politics with things" 109, 110
practice
  and ethics of envisioning 7–8
  of looking 6–7
  privileging 3–4
  processes and technologies 5–6
  research approach 4–5
  and theoretical perspectives 1–3
productive arts 85–6
  citing 89–91
  sighting 86–8
  siting 88–9
  see also drawings; paintings
provocation, landscapes as 111–14, 130

racial images, Caribbean 24–6
Ranciere, J. 75, 77
Reckwitz, A. 174–5

Red Road estate, Glasgow 133–5, 133–52
  modernist visions 135–9
  security and surveillance 145–6
  see also windows
Red Road (film) 147
Revival Field (Chin) 61
Rose, G. 4, 75, 94, 113, 162, 163
Royal Academy, London 81–2
rubbing, literal drawings and 92–5
Russell, B. 59

St Stephen's Hall, London 78–80
Salgado, S. 65–6, 69, 70–71, *72*
satellite/space technology 29–34
sculptures 61
security and surveillance 145–6
sensations/aesthetics 88–9
Serres, M. 6, 59, 60, 77
  and Hallward, P. 61
shipbreaking 62–7, 69–71, *72*
shipping
  Alcoa Steamship Company 23–8
  containers 49
Shooting into the Corner (Kapoor) 81–3
Shove, E. *see* Hand, M.
sighting (productive arts) 86–8
Simmel, G. 172, 182
Simms, M. 100, 103
siting (productive arts) 88–9
Smithson, R. 60–61
Smoke (film) 176–8, 179, 180
social-material practice 174–5
Solnit, R. 63, 67
Sontag, S. 61, 162
Southam, J. 62
space/satellite technology 29–34
Stallabrass, J. 65, 66
State Britain (Wallinger) 80–81
steel 60
  see also shipbreaking
steel-framed buildings 149
steel-framed windows 141, 149
Stephens, E.H. 31, 31*n*
  and Stephens, J. 30
stonework 95–7
Strachan, T. 32–4, *35*
suffragettes 80
surveillance, security and 145–6